Problem Solving Connections

Dr. William Driscoll
Donald W. Robb

Teacher Manual — Gold Level

Charlesbridge

CONTENTS

Introduction		iii
Unit 1	Strategies for Solving Problems	1
Unit 2	STRATEGY: Draw a Picture or Diagram	13
Unit 3	STRATEGY: Make a List, Table, or Chart	26
Unit 4	STRATEGY: Guess and Check	39
Unit 5	STRATEGY: Use Logical Reasoning	51
Unit 6	STRATEGY: Set Up an Equation	69
Expert Problems		80

The authors wish to express special thanks to Mj Terry of the Hartford Public Schools for her contributions to this project.

Copyright © 1992 by Charlesbridge Publishing. All rights reserved. This book may not be reproduced in any manner without the written permission of the publisher.

Publisher: **Charlesbridge Publishing**
85 Main Street, Watertown, MA 02172

Printed in the United States of America.

ISBN: 0-88106-657-5

10 9 8 7 6 5 4 3 2

INTRODUCTION

"When problem solving becomes an integral part of classroom instruction and children experience success in solving problems, they gain confidence in doing mathematics and develop persevering and inquiring minds. They also grow in the ability to communicate mathematically and use higher-level thinking processes."

— <u>NCTM STANDARDS</u> (1989), p. 23

It is with these same ideas in mind that *Problem Solving Connections* was created. The goal of this program is to encourage students to become confident problem solvers. To that end, *Problem Solving Connections* provides two essential elements:
1. a stimulating variety of problem types and
2. practical instruction in a range of problem-solving strategies.

This combination of elements is predicated on the belief that students need not only the *opportunity* to solve inherently interesting and challenging problems, but also *guidance* in productive techniques for seeking solutions. Problem solving strategies are not formulas to be memorized. Rather, they are ways of thinking to be modeled, discussed, and applied.

Often, there is *not* "a right way" to solve a problem. Instead, there are a number of critical thinking processes, any one of which may lead to a solution.

Some students may develop a range of these critical thinking processes independently and spontaneously. Other students develop these thinking processes only after observation, modeling, and discussion in very concrete situations. For this reason, *Problem Solving Connections* is organized around a number of specific strategies, which are introduced and elaborated at each level of the program. Students will have ample opportunity to try out each strategy, talk about how it works, and apply it to a rich assortment of problem situations.

STRATEGY INTRODUCTION

Problem Solving Connections is designed for use by students of all abilities in grades 3 through 6. To keep the focus on problem solving rather than on computational skills, problems for each grade level have been crafted to involve only those computational skills typically mastered by most children at that grade level, as shown in the following table.

LEVEL	GRADE	SKILLS
Orange	Grade 3	numeration, addition, subtraction
Blue	Grade 4	numeration, addition, subtraction, multiplication
Tan	Grade 5	numeration, addition, subtraction, multiplication, division
Gold	Grade 6	numeration, addition, subtraction, multiplication, division, fractions

Each book is comprised of six instructional units organized around specific problem solving strategies. These strategies are of two types: **supporting strategies** and **principal strategies**.

Supporting strategies are useful with a wide variety of problems. They are processes which, once learned, will be used in combination with the strategies in the following units.

Among the supporting strategies developed in this program are the following:

Estimate — arrive at an approximate answer, establish the range of possible answers, and check the reasonableness of an answer.

Act It Out — physically recreate the situation described in a problem so that a solution can be visualized.

Make a Model — use objects to represent the problem.

Solve a Simpler Problem — find a simpler problem to represent a more complex problem. As the simpler problem is solved, it provides a model of how to solve the more complex problem.

Work Backwards — a postponing strategy that allows the problem solver to ignore information which cannot yet be used and to concentrate on working systematically from the known to the unknown.

Supporting strategies are introduced in the first unit of each book of *Problem Solving Connections* so that they are available to students throughout the year.

Principal strategies are those which are the focus of later units. Once the problem solver has asked "What do I know?" and "What do I need to find out?" it is important for him or her to ask, "Which strategy would help me to solve this particular problem?" The following principal strategies are introduced in the Orange Level (Grade 3) and elaborated on at succeeding levels:

Guess and Check — for problems in which the given data allows you to make an educated guess at a solution and to use each successive guess, even though incorrect, to come closer to the solution.

Draw a Picture — for problems in which the verbal context can be more clearly seen in a picture or diagram. Making such a drawing will also result in a solution to the problem.

Make a List — for problems in which systematic organization of data into lists, tables, or charts leads to a solution.

Look for a Pattern — for problems in which the available data falls into clearly observable and definable patterns which can be used to predict a solution.

Use Logical Reasoning (Grade 4) — for problems in which making inferences based on an evaluation of available data leads to a solution.

Set Up an Equation (Grade 6) — for problems in which the most efficient way to organize the data is to use an algebraic equation.

SCOPE AND SEQUENCE

The following chart indicates the strategies associated with each unit in *Problem Solving Connections*.

The nature of the program and the elaboration of supporting strategies at the beginning of each level make it possible for students to use *Problem Solving Connections*, regardless of prior problem solving experiences. Teachers may comfortably use whichever book is appropriate to the skill level of their students (the Blue Level with advanced students in grade 3, for example, or the Tan Level with less advanced sixth grade students).

	UNIT 1	UNIT 2	UNIT 3	UNIT 4	UNIT 5	UNIT 6
ORANGE LEVEL (Grade 3)	Introduction to Problem Solving	Supporting Strategies	Guess and Check	Draw a Picture	Make a List	Look for a Pattern
BLUE LEVEL (Grade 4)	Supporting Strategies	Draw a Picture	Make a List or Table	Look for a Pattern	Guess and Check	Use Logical Reasoning
TAN LEVEL (Grade 5)	Supporting Strategies	Draw a Picture or Diagram	Make a List, Table, or Chart	Look for a Pattern	Guess and Check	Use Logical Reasoning
GOLD LEVEL (Grade 6)	Supporting Strategies	Draw a Picture or Diagram	Make a List, Table, or Chart	Guess and Check	Use Logical Reasoning	Set Up an Equation

Introduction

CURRICULUM CONNECTIONS

Problems in the program are grouped around curriculum topics so that they relate problem solving not only to mathematics, but to other subject areas as well. In this way, the critical-thinking strategies developed in these units can be seen as useful processes for language arts, science, social studies, and other subject areas.

Thus, problem solving can become part of a cross-curricular thematic unit, or extend a topic in any subject area at the teacher's discretion.

ORGANIZATION

At each grade level, *Problem Solving Connections* consists of a Teacher's Manual and a Student Book.

Student books contain six instructional units. Each unit is divided into five activities which develop and apply the strategy of the unit.

Each activity focuses on a particular aspect of the strategy being developed. The activities of a unit are therefore sequential and should be used in order. Each activity is generally related to a particular subject area. They are arranged in order of difficulty.

To help students get started, the unit begins with a summary or overview of the strategy being presented. In addition, the first problem of each activity contains a prompt in the form of a question to help students organize their thinking.

The last activity in each unit gives students an opportunity to consolidate and apply what they have learned. This activity may be used as an assessment of students' understanding, or as an optional activity for pairs or small groups of students.

In the Teacher's Manual, each unit contains an overview of the strategy, a list of suggested objectives, and specific teaching suggestions for each activity. In general, more suggestions are provided for earlier activities than for later ones.

Solutions for all problems appear in the manual alongside reduced copies of each student page.

Problem Solving Connections does not require that teachers be long-experienced or well-versed in problem solving techniques. Instead, the program seeks to make problem solving a joint learning experience for teacher and students alike.

PARENT INVOLVEMENT

Each unit of *Problem Solving Connections* contains a parent involvement page in the student book called "For You and Your Family." This page presents a nontechnical summary of the unit strategy along with several problems which students and parents can complete together.

This page may be torn out and sent home once the students have completed Activity 4 of any unit. Encourage students to return their papers the following day so they can share their solutions with the class.

EXTENSIONS

In addition to the problems contained in every unit, there is an additional section of cumulative review problems at the end of the book. These problems may be used at the end of each unit to maintain and reinforce problem solving strategies. They may also be used to assess students' understanding.

CALCULATORS / MANIPULATIVES

Suggestions for incorporating the use of calculators and manipulatives into classroom instruction are included whenever appropriate in the Teacher's Manual.

Introducing Cooperative Learning — If students are being organized into cooperative learning groups for the first time, you will want to spend some time explaining to them how the system works. One way to begin is to assign the teams routine tasks around the classroom to establish the concept of working together.

Another successful technique is to establish rules of operation:

1. Take responsibility for yourself. You are responsible for what is required of you in your team.

2. Take responsibility for your team. Work together and give everyone a chance to participate. Settle your disagreements among yourselves.

3. Reach agreement and understanding. Be sure everyone on the team understands and agrees with the team solution.

4. Answer your own questions. Only ask questions of the teacher if no one on the team can suggest an answer, or if all team members agree that the question is important enough to ask.

Team Roles — Teams sometimes work better when specific roles are assigned to each of the team members.

- The LEADER — states tasks, assigns responsibilities, and keeps discussion going.
- The RECORDER — takes notes as discussion progresses, and summarizes decisions and answers.
- The ENCOURAGER — makes sure that everyone is given a chance to participate, so that no one student dominates and no student loses interest.
- The MONITOR — makes sure that discussion and conversation remains relevant so that everyone keeps to the task at hand.

Rotate roles often so that students experience a variety of responsibilities. The more frequently students work in cooperative learning teams, the sooner they will become used to the dynamics of group work.

Classroom Organization — To encourage discussion and interaction, seat students at tables, or ask them to pull their chairs or desks close together in a square or circle.

Student interaction will result in some degree of noise. Students will need to talk among themselves as they seek answers to problems. As long as the conversation is productive and orderly, the volume will be tolerable.

Starting on a Problem — Many of the activities in *Problem Solving Connections* work best with whole-class discussion and initial instruction. However, even these discussions may involve the cooperative learning teams. Whole-class discussion may be interspersed with team assignments.

In some cases, you may want to assign all teams to work on the same problem. Each team can then report both its solution and the process it used to arrive at a solution. Other activities may involve having different teams work on different problems. Time available and teacher preference will determine how problems are assigned.

Throughout this Teacher's Manual, teaching notes indicate whether a problem is suitable for immediate assignment to teams or whether some degree of instruction should precede the independent team processing.

Reporting Results — When teams have reached a solution to a problem, have them report their answers to the class. If different teams report different answers, the class can compare and evaluate the various responses. In every case, the answer is no more important than the process by which the team arrived at the answer.

CLASSROOM SUGGESTIONS

Problem Solving Connections is a flexible program, intended to fit the needs of individual classroom teachers. Experience, however, does suggest some general guidelines.

1. *Problem Solving Connections* can be used equally well with a whole class or with small groups of students. In either case, teacher instruction should precede independent student practice. An important part of the instructional focus occurs *after* students have arrived at a solution, when they discuss together *how* they reached their solution.

2. Units should not be assigned for independent student work without teacher instruction unless students have had prior experience with problem solving. Students working alone, without appropriate instruction, may simply become frustrated.

3. After doing the first problem as an example with the whole class, many teachers assign the remaining problems to small groups or student teams that work cooperatively for a designated time.

4. Units may be taught as a block, taking perhaps a unit of time in the math class. Some teachers, however, periodically select one activity for class work. Others adopt the problem-a-day approach. Any of these approaches is entirely appropriate with *Problem Solving Connections*.

PROBLEM SOLVING AND COOPERATIVE LEARNING

Teachers across the country have long appreciated the effectiveness of cooperative learning. Problem-solving activities present ideal cooperative learning situations because they require the conditions associated with cooperative learning — a high degree of student interaction, peer motivation, and divergent thinking.

For teachers using *Problem Solving Connections* in a cooperative learning situation, we offer the following suggestions:

Forming Teams — Organize cooperative learning teams in groups of four. This number of students allows for productive interaction. Groups of three tend to decompose into a working pair and an isolated individual, while groups of five or more may develop into a leader and followers.

When the classroom does not divide evenly into groups of four, the extra students may be assigned to a group of four, making one or more groups of five. This approach seems to work better than a group of three.

Assigning Students to Teams — Research indicates that teams of mixed ability — two students of average ability, one of high achievement, and one of low achievement — work best. Such teams result in maximum interaction, the greatest gains in performance, and increases in self-esteem.

Problem solving is not sequential, nor does successful problem solving depend on prior mastery of mathematical operations. Therefore, a mix of students that is organized to foster interaction and cooperation provides a framework for exchanging points of view, generating alternative procedures, asking and answering questions, and stimulating a variety of contributions to the group activity.

In addition to a mix of abilities, each cooperative learning team ought to reflect the general composition of the classroom as a whole. If the class is more or less evenly divided between boys and girls, the composition of the teams should reflect the distribution. Similarly, if the class is composed of 25 percent minority students, each team should include a minority student. Mainstreamed special education students should be assigned to various teams, not be segregated into a team of their own.

A good technique for rotating the responsibility of reporting solutions and processes is to assign each team member a number (one, two, three, or four). Call on all the number one's to report on one problem, the two's on the next problem, and so on.

The Teacher as Monitor — Move about the room as the team members work together, observing how students in different teams are interacting. Encourage participation and help the teams to stay with the problem at hand.

Listening to what is happening in the groups will give you a sense of how to pace instructional activities as well as how to help team members function together.

PROBLEM SOLVING CONNECTIONS AND TAI MATHEMATICS

For teachers already using *TAI Mathematics*, these problem-solving activities make an ideal complement to the math program. Most TAI teachers have found that the "fourth week" approach provides the best format for incorporating problem solving into the classroom. Teachers spend about three weeks working with their teams and small instructional groups on the regular TAI materials; then, the fourth week, the class turns to *Problem Solving Connections*.

Each unit of *Problem Solving Connections* will take approximately one week. This schedule allows one class period for each of the four activities of the unit, plus one day for the Challenge Activity, which may be used as an assessment.

Unit Assignments — Since all students will be working on the same problem-solving unit, there is no need for a placement test. Units have been developed so that the student's operational skill level is not the only factor involved in successful problem solving.

Therefore, TAI classes should select the level of *Problem Solving Connections* that is appropriate to the grade level of the students: Orange for third grade, Blue for fourth, Tan for fifth, and Gold for sixth. The six units of each grade level will provide six weeks of problem-solving instruction and practice.

Use of Teams — For problem-solving activities, use the same teams that you have already set up for TAI. The only difference for the teams is that during problem-solving weeks, all members of the team will be working on the same assignment.

Teams should understand that it is the task of the team to find a solution to each problem assigned. All team members should contribute to finding a solution, and every team member should be able to explain the team's solution to the rest of the class.

Instruction — Work with the whole class during instruction sessions since all students will be working on the same types of problems. A useful technique during this period of whole-class instruction is to ask questions and give teams a minute or two to decide on an answer. In that way, instruction involves the teams actively from the very outset.

Once the first problem and its solution have been thoroughly discussed and students seem to understand both the problem and the solution, the next step will be for each team to solve the remaining problems of the activity.

Occasionally, some of the remaining problems may need additional comment from the teacher before student teams begin work on them. Those problems are pointed out in the Teacher's Manual as they occur.

Team Practice — Teams are expected to practice with a particular strategy only after they have received some instructions from the teacher. A team may choose to use an entirely different strategy to solve a problem and should actually be encouraged to choose any strategy that works for them. The point is that for most teams, problem solving experiments will be

Introduction

more satisfying and productive if some instruction precedes the team practice.

Assessment — The Challenge Activity (the last activity of each unit) may be used to assess students' understanding of the concepts taught in the unit.

You may also want to use the Expert Problems found at the back of the student book as an assessment. However, the strategies used here are cumulative, rather than unit specific. Problem solving strategy development is cumulative — that is, each experience in problem solving adds to the student's repertoire of internalized strategies. Therefore, there is no system of formal assessment in these units. Students may progress to the next unit of problem solving regardless of whether or not they have taken an assessment because problem solving is not a question of "mastery" of any one skill or operation.

The Challenge Activity, therefore, provides an option for evaluating how well a student has internalized a specific strategy. Although it provides useful information about a student, it need not prevent the student from exposure to future problem solving experiences.

Team Scores — If the Challenge Activities are used as an assessment, scores may be added to the TAI team scoring sheets. There are two possible approaches.

One is to enter the score for each student and add that to the scores for individual team members. Since all the Challenge Activities have five items, you might score 0 to 5 points for each team member depending on the number of acceptable solutions.

A second approach is to assign the Challenge Activity to the whole team and record not individual scores but a team total score.

UNIT 1

Strategies for Solving Problems

INTRODUCTION

For students whose problem-solving experience is limited, this unit will provide a foundation of highly productive strategies which will be useful along with the strategies taught in later units. For other students, the unit will provide a useful review of basic problem-solving strategies.

In addition, the unit activates the students' prior knowledge about problem types.

The unit focuses on a group of *supporting strategies* — so broad in their application that they are useful with many different problem types throughout the year. Five common strategies are included in this unit:

1. **Estimating** — determining the range of possible answers, arriving at an approximate solution, and checking the reasonableness of answers.

2. **Acting It Out** — performing the actions described in the problem so that potential solutions become more readily visible.

3. **Making a Model** — visualizing the conditions of the problem through the use of objects to represent details.

4. **Solving a Simpler Problem** — finding a simpler problem within a more complex problem. A solution to the simpler problem leads logically to a solution to the more complex problem.

5. **Working Backwards** — reading through the entire problem, deciding what information to use first, and working from the simpler problem to the more complex solution.

As students arrive at solutions to the problems in this unit, ask them to describe the strategies that they used.

OBJECTIVES

Students will
- distinguish among number, word, and process problems.
- convert number problems to word problems.
- create original process problems.
- use supporting strategies to solve problems.
- identify characteristics of five supporting strategies.

OVERVIEW

Activity 1 Identifying number, word, and process problems; estimating.

Activity 2 Converting word problems to number problems; creating process problems.

Activity 3 Using *Acting It Out* and *Making a Model*.

Activity 4 Using *Solving a Simpler Problem* and *Working Backwards*.

Activity 5 Identifying and defining various strategies.

Solving Problems
Student Book, pages 3 – 4

FOCUS: Identifying and solving word problems; estimating approximate answers

The problems in this activity are designed to lead students to an understanding of the differences between number problems, word problems, and process problems. Process problems — those which require us to develop a unique process of solution — will be the focus of the remaining activities of the unit and of the year's work in problem solving.

Distinguishing among problem types will help students to recognize similarities as well as differences. They will also learn some very fundamental questions to ask themselves before they solve a problem, regardless of type.

Introduce the concept of problem types by asking students to look at Problem 1 on student book page 3. Before they solve any of these problems, ask them first to estimate an answer. You many want to list several proposed estimates on the board. Students should also explain how they arrived at their estimates. When the problem has been solved, refer to the estimates and let students determine which was the closest.

As students begin to solve these problems, ask, **What do you need to know?** Apply this question to each of the parts of Problem 1. (For Problem 1a, they need to know the sum; in 1b, the difference; in 1c, the product; in 1d, the answer and the remainder.)

When they have solved all four problems, ask, **How did you know what to do in each problem?** (We used the signs.) **What operations did you use?** (addition, subtraction, multiplication, division)

As the students look at Problem 2, ask, **How are these problems different from the ones we just did?** (There are no signs to tell us what to do. There are words as well as numbers.)

Ask the students to read Problem 2a. Then ask, **What do you need to know?** (how many pencils were needed in all) **What do you already know?** (Each student needs 2; there were 27 students.) **What operation should you use?** (multiplication)

Use a similar procedure with the remaining problems. Before solving each one, students should first estimate approximate answers.

Students may note that Problem 2a is a one-step problem while 2b and 2c are multi-step.

When they have competed Problem 2, ask the students to describe the difference between number problems (Problem 1) and word problems (Problem 2). Possible answers: word problems refer to real things (students, pencils, tickets, books, etc.); number problems have signs; in word problems you have to figure out what operation (or operations) to use.

After students have read Problem 3, ask, **How is this problem different from the others?** (It's a word problem, but we can't just add, subtract, multiply, or divide.) **What do you need to know?** (how many combinations of two team members you can make) **Does anyone have an idea how to solve this problem?** Discuss student suggestions, and lead them to the conclusion that a list might help.

Suppose we call the members A, B, C, D, E, and F. Could we list all the combinations with A? (AB, AC, AD, AE, and AF) **How about combinations with B?** (BA . . .) **Haven't we already listed AB? Isn't that the same?** (yes) **So, what new combinations can we make with B?** (BC, BD, BE, and BF)

Have students continue in this manner until the problem is solved. For classes of more experienced problem solvers, simply assign the problem to individuals or teams and let them discuss their solutions.

Use the same questions — **What do we need to know? What do we already know? What process should we use?** — for Problems 4 and 5.

In Problem 4, students will discover that they can solve one part of the problem (who is the oldest? — Meg) but not the other (date of birthdays). The problem has both extra information (how old the girls are) and missing information (any one of the birth dates).

Suppose we know that Jennifer's birthday is September 16; can we solve the problem? (Yes, we can then add and subtract to find the other birthdays.) **What are they?** (Meg, Sept. 11; Edwina, Sept. 25)

Problem 5 also has no specific answer since we do not know exactly which coins he took out. We can determine the range of possible answers. ($6.60 left if he took out five quarters; $7.60 if he took out five nickels.) Now we have an estimate of some figure between $7.60 and $6.60. To determine exactly which coins he took out, extend the problem by suggesting that he had $7.30 left. Ask students to find the five coins that were left. They will need to make up a list of all possible coin combinations to see which one will leave exactly $7.30 in the bank.

Extension: Ask students to look up the definition of *strategy* and explain how it applies to what they did with Problems 3, 4, and 5.

SOLUTIONS

1. a. 6,609   b. 129   c. 12,896   d. 51 r 34

2. a. 54 pencils (27 x 2 = 54)

   b. 87 tickets

   $ 90.00 $217.50 ÷ $2.50 = 87
   + 127.50
   $ 217.50

   c. $ 8.08 left

   $ 21.00 $ 8.99 $ 17.98 $ 51.00
   + 30.00   x   2   4.95   − 42.92
   $ 51.00 $ 17.98   + 19.99 $ 8.08
                          $ 42.92

Unit 1 • Activity 1

3. 15 possibilities

4. Order of girls' birthdays: Meg, Jennifer, Edwina
   Birth dates: no solution. (Missing information is the date of any one birthday.)

5. Range: $6.60 to $7.60
   For example, if he has $7.30 left, he took out 1 q, 2 d, 2 n for a total of $ .55.

---

3. There are six members of the Math Team. In the annual competition, two members of the team will compete in the first event. How many different combinations of team members could the Coach use in the first event?
   THINK: How is this problem different from the ones you just finished? What question should you ask yourself?

4. Three friends all have birthdays in September. All three are 12 years old. Meg's birthday is five days before Jennifer's; Edwina's birthday comes exactly two weeks after Meg's. Who is the oldest? What is the date of each girl's birthday?
   THINK: Do you have all the information you need to solve the problem? What can you find using the given information?

5. Chan has been saving nickels, dimes, and quarters. He now has a total of $7.85 in his bank. If he took out five coins, how much money would be left?
   THINK: What is the most he could have left? What is the least? What information do you need to solve the problem?

I've Got a Problem
Student Book, pages 5 – 6

FOCUS: Distinguishing problem types; creating problems

By creating their own problems, students develop a deeper understanding of number, word, and process problems. In this activity, students will convert number problems into word problems (Problems 1 and 2) and create original process problems.

For Problems 1-3, the possible problems are unlimited. Each problem students create, however, should match the information given — same numbers, same operation. As an extension, students might like to create other similar problems to share with the class.

Before starting Problems 4 and 5, review the term *process problem* — a problem which requires a strategy to find a solution, rather than a computation.

Problem 4 lends itself to a *rank-order* problem ("Cal was born 2 years before Peggy . . .") or a *guess and check* type ("Together, the ages of Cal, Peggy, and Alicia add up to . . .").

Problem 5 will probably inspire problems of the *combination* type ("How many different combinations of events . . .").

The goal is for students to try writing original problems. You might want to offer a prize for the problem the class considers best, most unusual, or most interesting.

SOLUTIONS

Solutions will vary but should conform to the information given in each case.

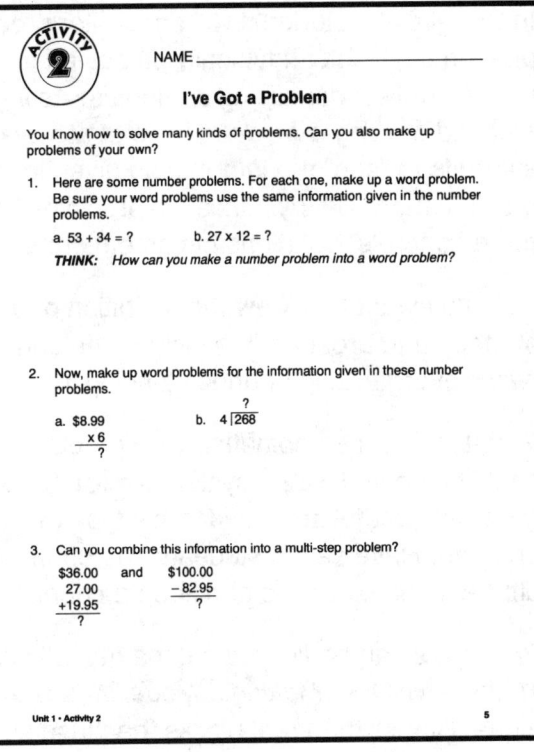

Unit 1 • Activity 2

5

Toys and Games
Student Book, pages 7 – 8

FOCUS: Supporting strategies — Acting it out, or making a model

In this activity, students will encounter process problems which can typically be solved through the use of two visualizing strategies: *Acting It Out* and *Making a Model*. These strategies help students to "see" the information given in the problem more clearly. Once the problems become more concrete, solutions become more evident.

You may want to review the definition of a strategy and broaden it to include the concept of "ways of organizing or understanding information."

Problem 1 is a combination-type problem. It could be solved with a systematic list, but *Acting It Out* is a useful strategy for this type of problem. Have seven students pair off in different combinations and keep a count.

Problem 2 can be figured out mathematically, but students will find *Making a Model* an easier technique. Rather than real blocks, because so many would be needed, the problem can be modeled as follows:

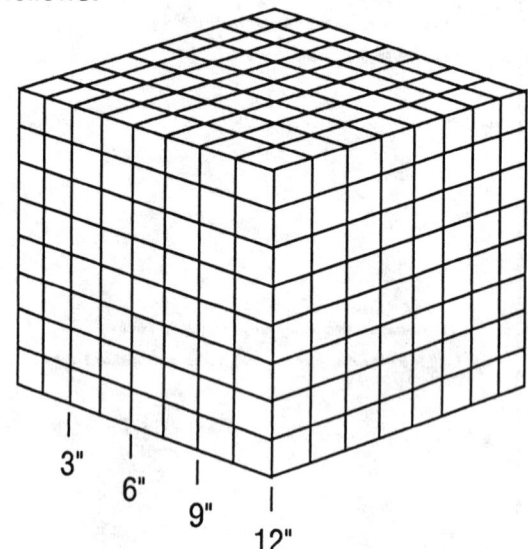

Once they construct this model, students will see that 8 x 8 produces 64 blocks in each of 8 layers, for a total of 512 blocks.

For Problem 3, let the students try to visualize a solution or construct their own model. If they have difficulty, suggest that they use the following layout to construct a model.

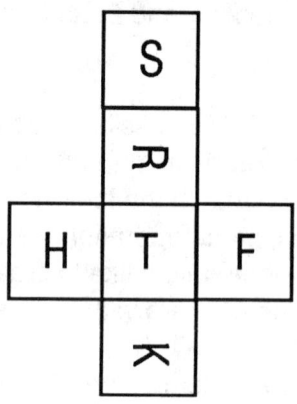

When they have made a cube, they can position it so as to check their solutions.

Problem 4 depends on understanding that a dropped ball will continue to bounce. A model will help students to visualize the information in the problem.

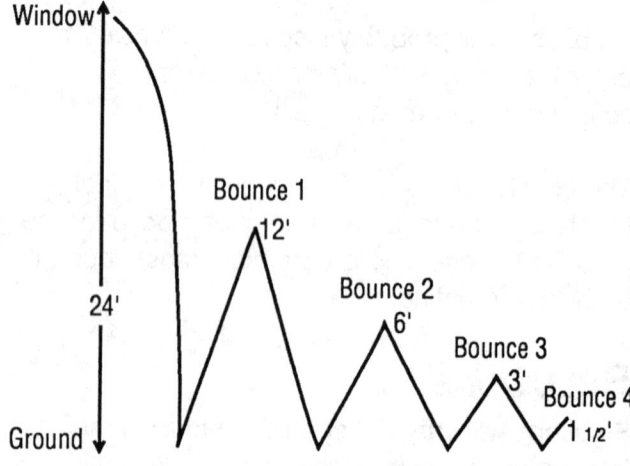

In Problem 5, students will discover that there are two routes they can take, each requiring them to land on 2 squares twice.

6

Unit 1 • Activity 3

SOLUTIONS

1. 21 games total; each plays 6 games
2. 512 blocks (8 x 8 x 8)
3.
4. She will catch it on the fourth bounce.
5. Yes, in eight jumps.

ACTIVITY 3 NAME ____________________

Toys and Games

1. The Chess Team has seven members. Each has to play a game against every other member. How many games will each member play? How many games will be played in all?

   THINK: *Could you and six classmates act out the problem?*

2. Larry's little brother loves blocks. He has hundreds of them. The blocks measure 1 1/2 inches high by 1 1/2 inches wide by 1 1/2 inches deep. How many blocks are in a stack 1 foot high by 1 foot wide by 1 foot deep?

3. The blocks have letters on each side. Here are several views of one block. Can you fill in the missing letters?

Unit 1 • Activity 3 7

4. Every time I drop my large rubber ball, it bounces up exactly 1/2 the distance from which I dropped it. Each time it bounces, it bounces up 1/2 the distance of the bounce before. I dropped it from the attic window down to the driveway, a distance of 24 feet. My sister caught it when it had bounced up 1 1/2 feet. How many times did the ball bounce?

5. Here is a checker board. You are the player with only the dark checker at the lower left. It has been kinged so it can move in any direction. Can you win the game in one move by jumping all your opponent's checkers?

8 Unit 1 • Activity 3

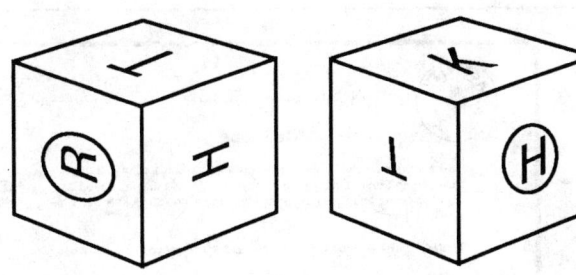

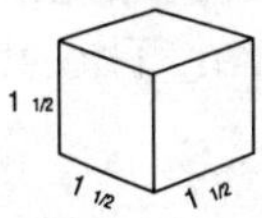

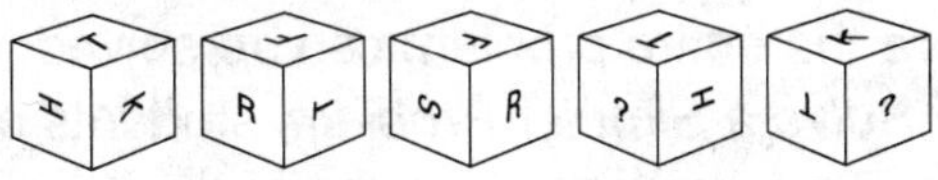

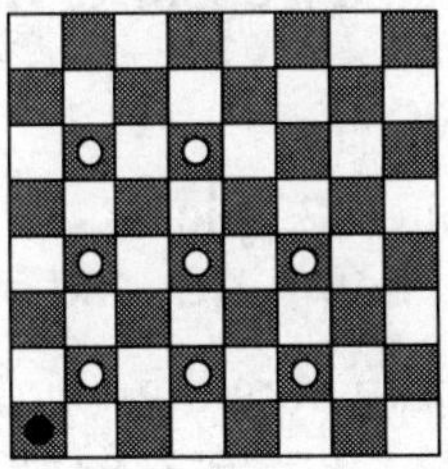

Odd Jobs
Student Book, pages 9 – 10

FOCUS: Working backwards and solving a simpler problem

In this activity, students will discover that sometimes we need to save information in a problem until later information makes it possible to use the information we passed over. The strategy is called *Working Backwards.*

They will also learn that some problems contain hidden, simpler problems which must be solved before the entire problem can be solved. Once they *Solve a Simpler Problem,* students can *Work Backwards* to solve the entire problem.

The problems in this activity are not difficult once students recognize the value of these strategies. Problems 1-3 involve *Working Backwards.* You may want to discuss this strategy as students attempt their solutions to the first problem. Ask, **Why did we have to ignore some information at first?** (because we couldn't use it until we had other information)

Help the students to discover the hidden problems in Problem 4 (How many square feet in each lawn? How long does it take to cut 2000 square feet?) and Problem 5 (How many boxes fit in one carton?) In both these problems, students may find it useful to *Make a Model.*

SOLUTIONS

1. Quon painted 35 pickets.
   Steve, 16; Lorna, 19; and Melinda, 38

2. 15 more hamsters than birds
   12 cats; 24 hamsters; 4 dogs; 9 birds

   ```
    24
   - 9
   ---
    15
   ```

3. 164 papers in all
   Claudia, 38; Mark, 30; Yvonne, 33; Sasha, 21; Paul, 42

4. 10 minutes
   60 x 100 = 6000 sq. ft. ÷ 200 sq. ft. per min.
   40 x 50 = 2000 sq. ft. ÷ 200 sq. ft. per min. = 10 min.

5. 8,640 boxes
   Each carton: 6 layers of 12 boxes each = 72 boxes
   72 x 120 = 8,640

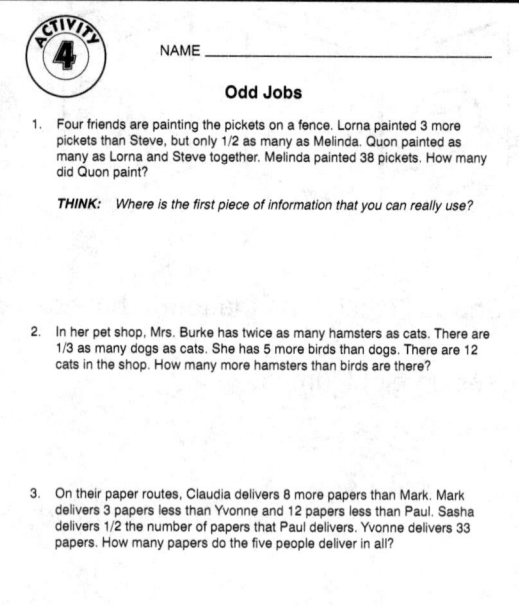

Unit 1 • Activity 4

Challenge
Student Book, pages 11 – 12

FOCUS: Review of supporting strategies

These problems can serve as an assessment of students' understanding. They could also be assigned to small groups of students as additional practice. You may want to assign tasks to each group member to promote a cooperative learning group setting.

Note: Since this unit introduces several supporting strategies, encourage students to discuss not only their solutions, but also what strategy they chose and why they chose it.

SOLUTIONS

1. 165 blocks in 5 layers

    9 x 9 = 81
    7 x 7 = 49
    5 x 5 = 25
    3 x 3 = 9
    1 x 1 = 1
    ─────────
          165

2. They have earned enough for one, but not for three. A rough estimate of 40 per hour times 9 hours = 360. Even if all were children, that would be $540. If all were adults, 360 x $3 = $1,080, not enough for three. Since some of those were probably children, there would not be enough even for two.

3. Julie, George, Pam, Shana, Rashid, Kumi

4. 72 in all
    32 apples, 16 oranges, 12 bananas, 8 plums, 4 pears

5. 38 feet long
    Simpler problem: How many bricks per foot?
    24 bricks per foot; 912 ÷ 24 = 38

 NAME _____

Challenge

1. I built a pyramid of blocks. The bottom layer was 9 rows by 9 rows. The second layer was 7 rows by 7 rows; the next layer, 5 rows by 5 rows and so on. How many layers did it take to build the pyramid? How many blocks did I use?

2. Tickets to the School Carnival are $3.00 for adults and $1.50 for children. The school hopes to earn money for new computers, which cost $500.00 each. Here are the numbers of admissions to the carnival today:

Time	Admissions
9:00 -10:00	36
10:00 -11:00	48
11:00 -12:00	50
12:00 - 1:00	56
1:00 - 2:00	26
2:00 - 3:00	32
3:00 - 4:00	44
4:00 - 5:00	46
5:00 - 6:00	40

    Without knowing how many of these were children and how many were adults, can you estimate to see if they have enough for a computer? How about three computers?

3. Six people are standing in line. George was not at either end. Two people were in line after Shana. No one was ahead of Julie. Shana was not next to George. Pam was ahead of Kumi and Rashid. There were three people between Julie and Rashid. What is the order of the line?

4. I bought fruit at the market for the class picnic. I bought
    • 3 times as many bananas as pears.
    • 1/2 as many plums as oranges.
    • As many oranges as pears and bananas together.
    • Twice as many apples as oranges.
    • 4 pears.

    How many pieces of fruit did I buy in all?

5. You are laying bricks in a path. The bricks are 6 inches by 3 inches. The path is 3 feet wide. To build the path, you need 912 bricks. How long is the path?

Unit 1 • Activity 5

STRATEGY DISCUSSION

Assign different groups or teams in the class to provide a definition of each of the five strategies introduced in the unit: *Estimating, Acting It Out, Making a Model, Solving a Simpler Problem,* and *Working Backwards.* Each team can present its definition to the class.

Ask students to comment on each of the definitions. Based on the class discussion, ask each team to produce a poster with a strategy definition for display on a bulletin board in the classroom.

The definitions presented on page 1 of this Teacher's Manual can serve as guidelines for the class definitions.

EXTENSION

Ask each team to create a problem (or problems) that can be solved using the strategy which the team has defined.

Strategies for Solving Problems
Student Book, pages 13 – 14

This page of the student text is designed to be taken home as a family involvement activity.

Included for the parent is a brief explanation of the unit content with a summary of the strategy which has been developed.

Along with that background information, there are several problems which the student and the family can solve together.

You may want to ask students to return these papers to school on the following day and share their solutions with the class.

This page may be assigned as soon as Activity 4 of this unit has been completed.

SOLUTIONS

1. $60   week 1 $60
                        2 $50
                        3 $25
                        4 $15

   Strategy:
   Work Backwards

NAME _____

Strategies for Solving Problems

Strategies are ways of thinking about the information we find in problems. In our math class, we have been learning about several strategies that make problem solving easier.

- *Estimating* means finding an approximate answer — not an exact answer — and checking to see that our answers make sense.
- *Acting It Out* and *Making a Model* are ways of "seeing" a problem clearly so we can find a solution.
- *Solving a Simpler Problem* means that sometimes we find an easier problem hidden within the problem. Once we find that solution, we can use it and *Work Backwards* to a solution for the entire problem.

These strategies will help us solve many of the problems we will explore this year. You may want to use the following problems at home to help your child understand them better. They are similar to problems we have done in school.

1. Every week that an item is not sold at the Bargain Basement Store, the manager reduces the price. One week, Mickey saw a sweater he liked. The next week, it was $10.00 less. The third week, the price was reduced by 1/2. The next week, the price went down another $10.00, and Mickey bought the sweater for $15.00. How much did the sweater cost the first week that Mickey saw it?

   SUGGESTION: Start with the final price and work back from there.

2. 7 trips
   1. Take rooster across.
   2. Return alone.
   3. Take cat or seeds across.
   4. Return with rooster.
   5. Take seeds or cat across.
   6. Return alone.
   7. Take rooster across.

   Strategy:
   Act It Out

3. Bill   6 hours @ $5.50 = $33.00
   Mike   4 hours   = $22.00
   Effie   3 hours   = $16.50
   Suellen 2 hours   = $11.00

   Strategy:
   Solve a Simpler Problem/Work Backwards

   Simpler Problems: How long did each person work? How much does the manager pay per hour?

---

2. My Aunt Mae was walking down a country road with her cat, Grover, and her pet rooster, Leonard. In a small basket, she was carrying a sack of sunflower seeds. She came to a wide creek that she would have to wade across. She could carry her pets and the sunflower seeds across in the basket, but only one thing would fit in the basket at a time.

   She knew she could not leave Grover alone with Leonard, even for a few minutes because the cat would chase the rooster. She could not leave Leonard with the seeds because he would eat them.

   To get the cat, the rooster, and the seeds across the creek, she had to make several trips. Can you decide how many trips she made, and what she carried on each trip?

   SUGGESTION: Use objects to represent the cat, rooster, and seeds and act out the problem.

3. It takes an hour to wash five windows at the Fairhaven Apartment Complex. Bill started washing at 10:00 AM on Saturday, and worked until noon. At 11:00 AM, Mike began helping him. They took an hour for lunch, starting again at 1:00 PM. Just then, Effie joined them. They worked until 3:00 PM when Suellen came along to help, too. All four worked for an hour until Mike and Effie had to leave. The others finished work at 5:00 PM.

   The manager paid a total of $82.50 for the window washing job. How much did each person earn?

   SUGGESTION: Think about how the manager could divide the money fairly.

UNIT 2

STRATEGY: Draw a Picture or Diagram

INTRODUCTION
The previous unit focused on supporting strategies, which are useful in a variety of problem solving situations. Supporting strategies can be used with a wide range of problem types.

This unit, and those which follow, will focus on principal strategies. This unit introduces the principal strategy *Draw a Picture or Diagram*. Use of a principal strategy is dictated by the nature of the problem itself. A principal strategy will work for some kinds of problems, such as those in this unit. Other strategies, specific to other types of problems, require strategies such as those introduced in later units.

Different problems in this unit will require different types of drawings or diagrams. All, however, are more readily solved once students have represented concretely the information they contain. Student drawings or diagrams should help them to —
- represent pictorially the complexities of the problem
- understand the problem more clearly
- avoid obvious errors
- visualize a solution.

OBJECTIVES
Students will
- create a picture or diagram to represent the conditions of a given problem.
- verbalize the relationship between their representation and the information in the problem.
- use pictures or diagrams to solve problems.
- use visualizing in combination with various supporting strategies.
- identify conditions in problems which make visualizing an effective strategy.

OVERVIEW
Activity 1 Visualizing geometric designs.
Activity 2 Recognizing geometric relationships.
Activity 3 Representing complex information in a problem.
Activity 4 Using a variety of representations based on problem conditions.
Activity 5 Review of various problem types.

When students have completed this unit, you may wish to assign the Expert Problems, Set A, on pages 85-86, for cumulative strategy application.

By Design
Student Book, pages 17 – 18

FOCUS: Creating geometric shapes
CONTENT CONNECTION: Area and perimeter (MATHEMATICS)

As an introduction to the strategy *Draw a Picture or Diagram,* these problems are designed to focus student attention on representing accurately the information and conditions of the problem.

Though the problems are not particularly difficult, they do assume prior knowledge of certain essential terms: *area, perimeter,* and *square feet.* All the problems involve using 1-foot-square tiles. You may want to review this concept by putting the following on the board:

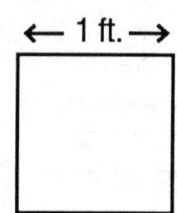

Ask, **If this is a square, how long are each of the sides?** (one foot) **What does "one square foot" mean?** (an area one foot long on each of its four sides)

Have the students look at the drawing again, and ask, **What does *perimeter* mean?** (the distance around the outside edge of the figure) **What is the perimeter of this figure?** (four feet)

Now add another square alongside the first one:

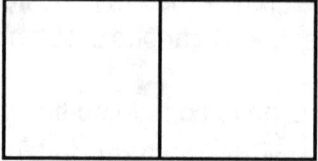

Ask students to calculate the area (two square feet) and the perimeter (six feet). Point out that since each square touches the other completely along one side, each square has only three outside edges, which are part of the perimeter.

Once students understand these basic concepts, they should be able to solve the problems independently.

MANIPULATIVES: *You may want to provide 1-inch squares of paper which students can move around. Graph paper could also be used.*

SOLUTIONS

1.

   or any combination of three tiles in one row and two in another

2. There are at least five ways:

   (These figures can of course be rotated, but

   and

   are essentially the same figure.)

3. smallest possible perimeter: 12 feet

   largest possible perimeter: 20 feet

4. Eight tiles can be arranged with perimeters of

   12 feet

   14 feet

   16 feet

   18 feet

ACTIVITY 1

NAME _____

By Design

Your job is to create designs using tiles that are 1 foot by 1 foot. In each design, every tile must touch at least one other tile completely along one side (like this: ☐☐ ; not like this: ☐◳ .)

1. Can you create a design that has an area of five square feet and a perimeter of ten feet?

   THINK: How many tiles will it take?
   How do you measure perimeter?
   How should you arrange the tiles?

2. How many different ways can you create a design that has an area of four square feet and a perimeter of 10 feet?

3. In a design whose area is nine square feet, what is the smallest possible perimeter? What is the largest?

4. In a design with an area of eight square feet, how many different perimeters are possible?

5. To create a design with a perimeter of 22 feet, how many tiles will you need?

---

5. 10 tiles

Unit 2 · Activity 1

Vexing Hexagons
Student Book, pages 19 – 20

FOCUS: Recognizing patterns in geometric elements
CONTENT CONNECTION: Designing patterns (ART)

The problems of this activity are designed to underscore the need for accurate inclusion in drawing the details given in the problem. They also demonstrate how one strategy can work in conjunction with another. The solution to Problem 1 becomes evident if students construct a drawing such as the one in the solutions. Students can count the 24 hexagons in that ring. They may also see the emergent pattern: an increase of six hexagons in each successive ring.

Once students have drawn the four-ring diagram for Problem 1, they can use their drawings as a basis for solving Problems 2 and 3. Ask, **What color was the first ring?** (black) **The third ring?** (black) **What color do you think the fifth ring will be?** (probably black) **Why?** (because odd-numbered rings are black)

Next, ask, **How many hexagons in the first ring?** (six) **How many in the second?** (twelve) **In the third?** (eighteen) **How many more hexagons in each extra ring?** (six)

Students now have all the information they need to solve Problem 2, even without using a drawing. Ask, **Is 15 an odd or an even number?** (odd) **Then what color are the tiles in row 15?** (black) **How did you know?** (The odd-numbered rings in the drawing have black tiles.)

To answer the remaining question in Problem 2, suggest that students organize the information they have into a chart:

Ring	White	Black
1		6
2	12	
3		18
4	24	

Ask, **What could we do now to solve the problem?** (complete the chart through ring 15)

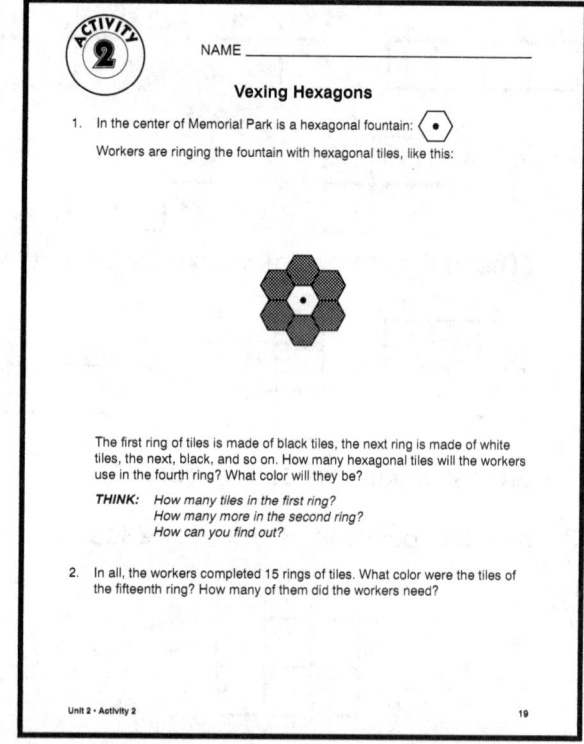

Once completed, the chart also supplies the information needed to solve Problem 3.

MANIPULATIVES: You may wish to provide students with paper hexagons instead of having them draw their own.

As in Activity 1, most students should be able to solve these problems independently.

SOLUTIONS

1. 24 white tiles (Each ring requires 6 tiles more than the previous ring.)

2. 90 black tiles

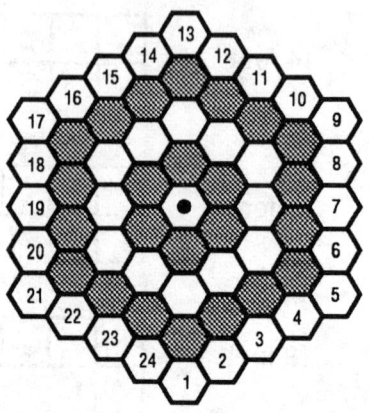

Unit 2 • Activity 2

3. More black tiles: 384

Ring	1	2	3	4	5	6	7	8	9	10	11	12	13	14	15	Total
Black	6		18		30		42		54		66		78		90	384
White		12		24		36		48		60		72		84		336

4. 3 rings   (1 + 6 + 12 + 18)

5. 4 trapezoids; 6 triangles

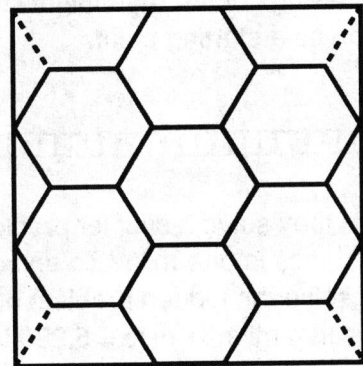

---

3. When they finished, had the workers used more white tiles or more black ones?

4. You are using small hexagonal tiles, all the same color, to cover a square table top. You center the first tile exactly and build up rings of tiles around it. When finished, you have used 37 tiles in all. How many complete rings have you made?

5. Marty and Tim cut some hexagons out of colored paper to cover a bulletin board. They used seven hexagons, and the bulletin board looked like this:

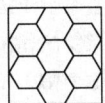

They also cut out some trapezoids (⬡) and some triangles (△). How many of each will they need to cover the bulletin board completely?

Unit 2 • Activity 2   17

Ancient Architecture
Student Book, pages 21 – 22

FOCUS: Keeping track of complex information
CONTENT CONNECTION: Ancient civilizations (SOCIAL STUDIES)

The problems of this activity are more complex than those solved earlier in the unit. Each problem contains a number of details which students will need to reflect accurately in their drawings. Students may need guidance in translating verbal descriptions into drawings.

MANIPULATIVES: *Graph paper would be useful for the drawings of this activity. It would be helpful to supply each student with 6 or 8 sheets of graph paper.*

To start, call on one student to draw on the board the basic shape of the temple described in Problem 1. (It is a rectangle longer than it is wide.) Students can then make their own drawing on graph paper.

Since 8 is a common denominator of both 96 and 56, suggest that they let each square on the graph represent 8 feet. Ask, **How many squares long will the drawing be?** (12; 12 x 8 = 96) **How wide?** (7 squares; 7 x 8 = 56) Check to see that students have drawn the basic plan correctly. Then ask, **How far from the walls are the pillars?** (8 feet) **How many squares?** (one square) Have students indicate the pillars with dots. **How far apart from each other should they be?** (8 feet, or 1 square)

Students should now have no difficulty counting the pillars to solve the problem.

Problem 2 contains several hidden problems:
- What are the dimensions of the paving blocks? (8 inches by 8 inches by 8 inches, or 1/3 of 2 feet, which equals 24 inches)
- How many blocks does it take to reach from edge to edge? (30; 30 x 8 = 240; 20 feet x 12 inches = 240)

Students' drawings of the road might show a cross section as a starting point.

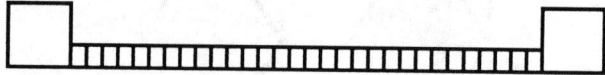

Suggest that they solve a simpler problem: How many stones in one mile? To do so, they will need to solve the hidden problem of how many inches in a mile. (1 mile = 5,280 feet; 5,280 x 12 = 63,360 inches) They can then divide by 8, the number of inches per stone, to get the number of rows. (7,920) Then they can multiply by the number of bricks in a row (30, already determined). Since 30 x 7,920 = 237,600, it takes 237,600 stones per mile.

Multiplying 237,600 by 26 (the number of miles) gives 6,177,600 — the solution to the problem.

CALCULATORS: *Since this problem requires a great deal of multiplying of large numbers, the use of calculators will facilitate finding the solution. Calculators would also help in Problems 3, 4, and 5.*

Problem 3 is fairly straightforward. Students will need only to calculate the number of blocks in each layer and keep a running total.

You can extend this problem by asking students to calculate how many blocks will be used in all. (3,268) **If the blocks are 6 feet by 6 feet, how many square feet does the pyramid cover?** (36,864) **How tall is it?** (54 feet) Note that students do not need to know the block dimensions to solve the original problem. That is simply extra information.

Let students try Problem 4 on their own. It contains two simpler problems to solve. How many pillars? (56) How many blocks per pillar? (78)

Problem 5 requires figuring out the planned dimensions of the pasture. We can multiply the number of posts by 4 to get the perimeter of the field (136 feet). Half the perimeter would therefore be 68 feet. Knowing half the perimeter gives us the total of the length and the width. Next we have to find two numbers whose sum is 68 and whose difference is 12. Using a *Guess and Check* strategy gives us 40 and 28. We have now solved the first set of hidden problems.

We now know that the pasture was planned to be 1,120 square feet (40 x 28). But the barn juts into the pasture, reducing the square footage. We know the farmer saved 4 posts. Therefore, since one must have been a corner post, the area of fence he cut out looked like this:

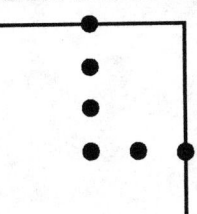

Since the posts are 4 feet apart, we can figure this area as 96 square feet (8 x 12).

If we subtract 96 from 1,120, the answer is 1,024 — the solution to the problem.

SOLUTIONS

1. 46 pillars
2. 6,177,600 stones

   As an extension, students might determine that the road also requires 137,280 edging stones.

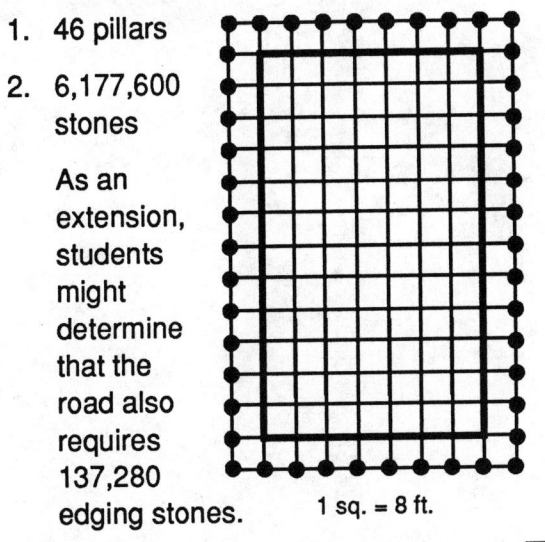

1 sq. = 8 ft.

 NAME _____

Ancient Architecture

1. The dimensions of one ancient Greek temple are 96 feet by 56 feet. Around the outside, eight feet from the temple walls, is a row of pillars. If the pillars are eight feet apart, how many are there?

   THINK: Can you draw a picture to show the temple walls and the location of the pillars?

2. The Romans were great road builders. One of their roads was edged with blocks of stone 2 feet wide by 2 feet high by 2 feet deep. Between the edging blocks, the road was 20 feet wide. The road was made of paving blocks that are 1/3 the dimensions of the edging blocks. How many paving blocks did they use to make the road 26 miles long?

3. The Mayans of Mexico and Central America built huge pyramids of solid stone blocks. The square base of one pyramid is 32 blocks by 32 blocks. The next layer is 28 blocks by 28 blocks, and so on. The blocks are 6 feet by 6 feet by 6 feet. Three layers have been laid. The top layer was supposed to be 2 blocks by 2 blocks. How many more blocks would they have needed to complete the job?

Unit 2 • Activity 3   21

3. 884 blocks

   First 3 rows
   ```
     576      3268
     784     -2384
   +1,024     ─────
   ──────      884
    2,384
   ```

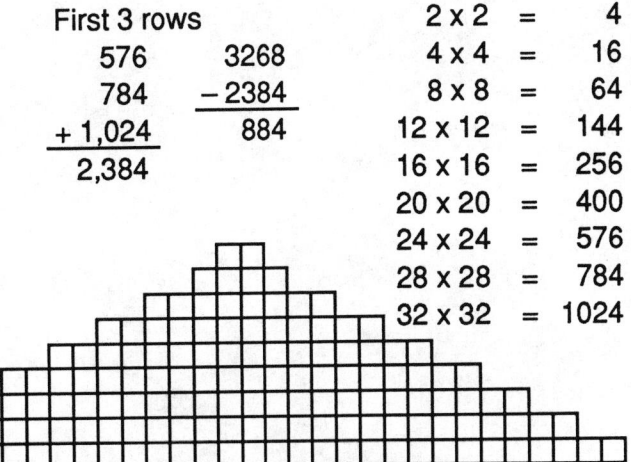

2 x 2	=	4
4 x 4	=	16
8 x 8	=	64
12 x 12	=	144
16 x 16	=	256
20 x 20	=	400
24 x 24	=	576
28 x 28	=	784
32 x 32	=	1024

Unit 2 • Activity 3   19

4. 4,368 blocks

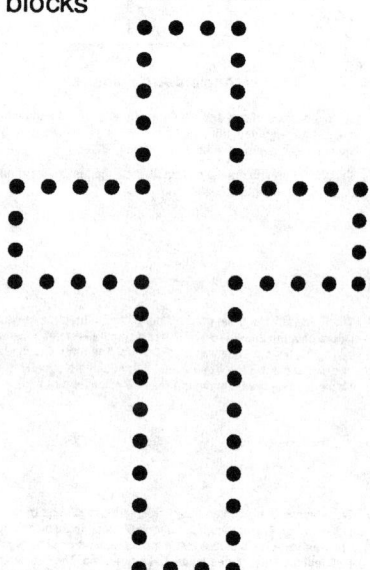

5. 1,024 square feet

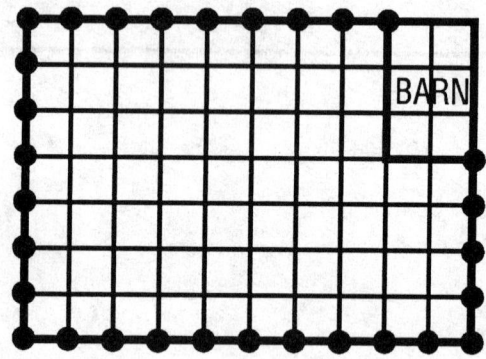

Put posts every 4 feet. 30 posts are needed if attached to Barn. 1,024 sq. ft.

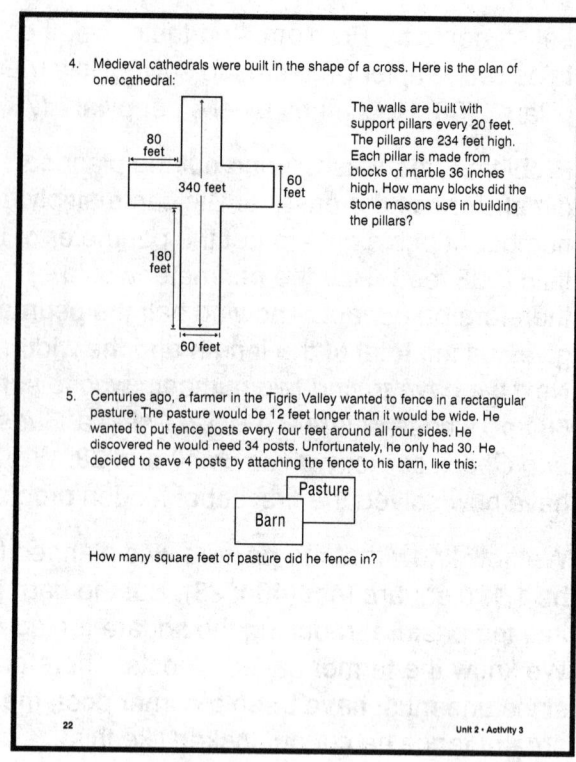

Unit 2 • Activity 3

Roughing It
Student Book, pages 23 – 24

FOCUS: Choosing an appropriate type of drawing
CONTENT CONNECTION: Camping out (OUTDOOR EDUCATION)

Each problem in this activity will require a different type of drawing to represent additively the conditions of the problem. Experienced problem solvers should have little difficulty with them. Students less familiar with problem solving strategies may require more assistance and teacher direction.

Problem 5 contains two simpler problems to be solved first: how many packets of stew will be needed (4; 3 will not feed 36 people) and how much water is needed? (4 liters)

Problem 5, like all "pouring" problems, is essentially a *Work Backwards* problem. Students will need to start at the end — with four liters in one container. It will have to be the 5-liter container since the other one will not hold four liters.

How can you get four liters? (Pour out one.)
How do you measure it? (It would have to go into the 3-liter container with 2 liters already there.) Working with various combinations such as 5 – 1 = 4, 2 + 1 = 3, 3 + 2 = 5 (always using sums that equal one of the containers), students can work back to a starting point.

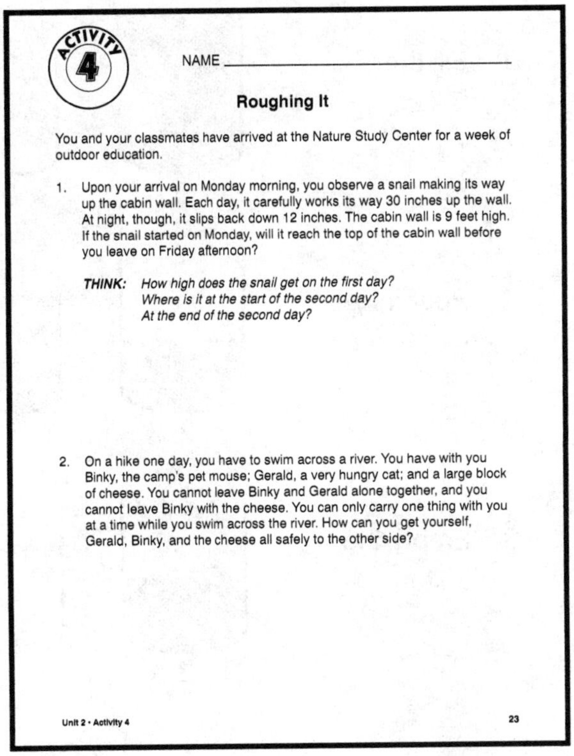

Roughing It

You and your classmates have arrived at the Nature Study Center for a week of outdoor education.

1. Upon your arrival on Monday morning, you observe a snail making its way up the cabin wall. Each day, it carefully works its way 30 inches up the wall. At night, though, it slips back down 12 inches. The cabin wall is 9 feet high. If the snail started on Monday, will it reach the top of the cabin wall before you leave on Friday afternoon?

    THINK: How high does the snail get on the first day?
    Where is it at the start of the second day?
    At the end of the second day?

2. On a hike one day, you have to swim across a river. You have with you Binky, the camp's pet mouse; Gerald, a very hungry cat; and a large block of cheese. You cannot leave Binky and Gerald alone together, and you cannot leave Binky with the cheese. You can only carry one thing with you at a time while you swim across the river. How can you get yourself, Gerald, Binky, and the cheese all safely to the other side?

Unit 2 • Activity 4   23

SOLUTIONS

1. No, the snail will not reach the top of the wall until Saturday.

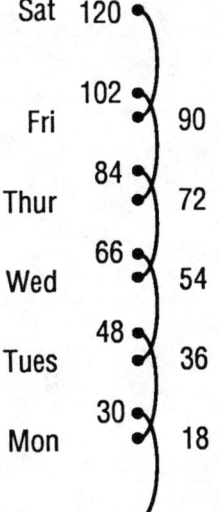

2. 7 trips:

	Left on Side 1		Left on Side 2
1.	Gerald cheese	Binky →	
2.	Gerald cheese	← nothing	Binky
3.	Gerald	cheese →	Binky
4.	Gerald	← Binky	cheese
5.	Binky	Gerald →	cheese
6.	Binky	← nothing	Gerald cheese
7.		Binky →	Gerald cheese

3. No, you cannot leave the dog alone with any of the other items.

4. Ito, followed by Nadine, Albert, Jessica, and Polly

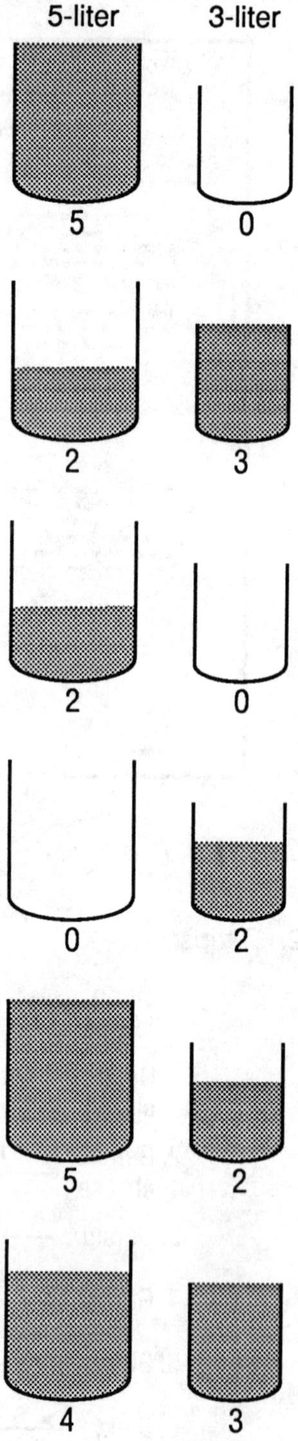

5. You will need 4 liters of water. Do it in 6 steps:

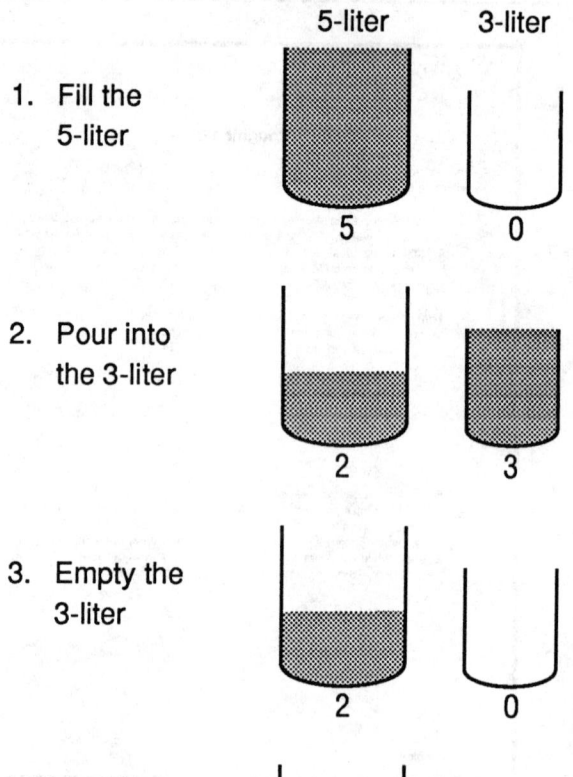

	5-liter	3-liter
1. Fill the 5-liter	5	0
2. Pour into the 3-liter	2	3
3. Empty the 3-liter	2	0
4. Pour into the 3-liter	0	2
5. Fill the 5-liter	5	2
6. Pour into the 3-liter	4	3

Challenge
Student Book, pages 25 – 26

FOCUS: Review of problem types

These problems can serve as an assessment of student's understanding. They could also be assigned to small groups of students as additional practice. You may want to assign tasks to each group member to promote a cooperative learning group setting.

SOLUTIONS

1. 18 trees will provide the 36 posts he needs.

2. You will arrive on the following Monday.

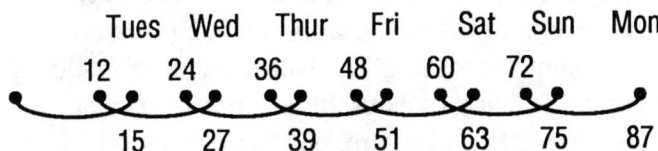

3. Yes

   One example: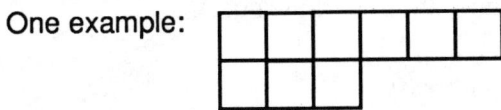

   There are several others.

Unit 2 • Activity 5

4. The order, starting with the most popular, is strawberry, chocolate, peanut butter fudge, mint chocolate chip, rocky road, and butter pecan.

5. 5 steps:

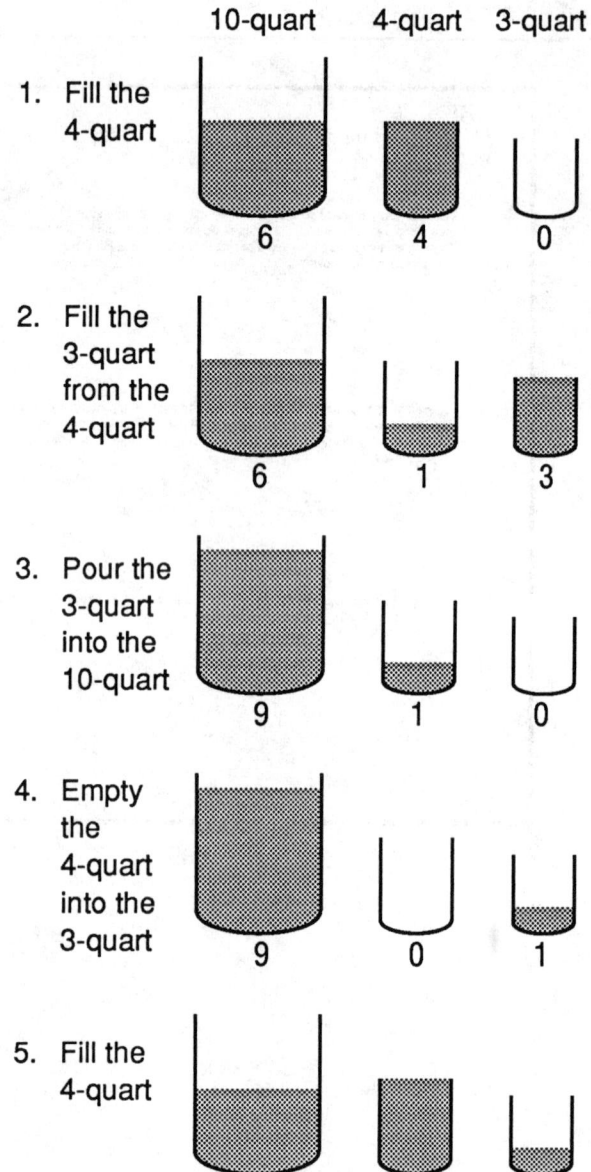

> 4. Luke took an ice-cream flavor survey. Among the six flavors on which he polled people, he discovered that chocolate was in second place. More people liked rocky road than butter pecan. Two flavors were more popular than peanut butter fudge. Strawberry was more popular than mint chocolate chip. Only one flavor was less popular than rocky road. Can you list the flavors in order of popularity?
>
> 5. You need 5 quarts of oil. You have a full ten-quart container. You also have a four-quart container and a three-quart container, both empty. Without wasting any oil, how can you measure out exactly five quarts?

Strategy Discussion

Review with the students the ways in which drawings can be a useful problem-solving strategy by making the problem clearer, helping to organize information, and avoiding errors.

Ask them to recall the kind of problems in which drawings or diagrams were helpful. (rank order problems, "crossing the river" problems, climbing up and sliding back problems, problems with fence posts and pillars, pouring problems)

See if students recognize any similarities among these problems. (The students may suggest that the problems were all easier to solve when they could "see" them in drawings.)

Extension

Several of the problems in this unit, such as the "crossing the river" problems or the climbing up and sliding back problems, are "classic puzzlers" that often appear in magazines. Students might like to make posters summarizing "rules" for solving puzzles like these. Posters could be illustrated and put on display.

Draw a Picture or Diagram
Student Book, pages 27 – 28

This page of the student text is designed to be taken home as a family involvement activity.

Included for the parent is a brief explanation of the unit content with a summary of the strategy which has been developed.

Along with that background information, there are several problems which the students can solve together with their families.

You may want to ask students to return their papers to school on the following day and share their solutions with the class.

This page may be assigned as soon as Activity 4 of this unit has been completed.

SOLUTIONS

1. Here are two possibilities. There are others.

2. The order is Doggy Dinner, Canine Chow, Puppy Wonder, Mutt Meals, Tasty Treats, and Yummy Bits.

3. 222 bricks. Each row of the three required rows will use 74 bricks.

   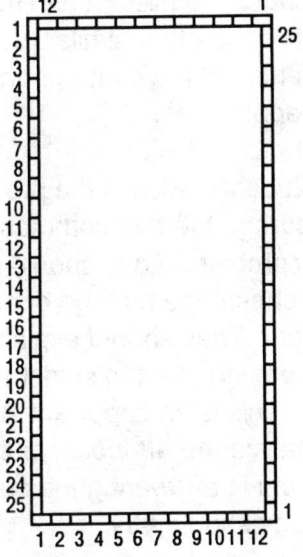

4. Six days.

   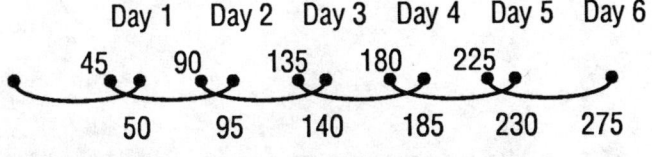

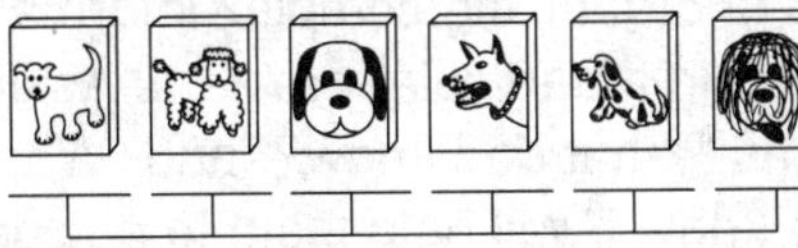

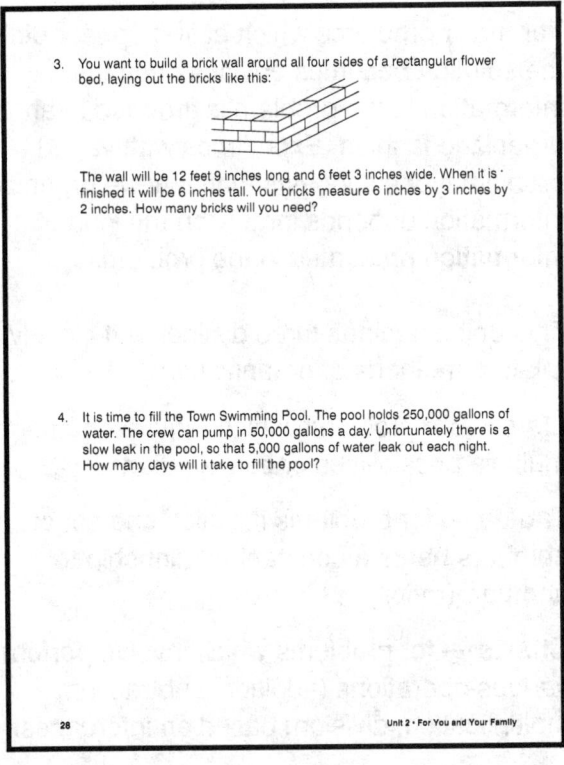

Unit 2 • For You and Your Family

25

UNIT 3

STRATEGY: Make a List, Table, or Chart

INTRODUCTION

In this unit, the students will need to select a method of organizing complex information contained in a particular problem. Asking the questions "What do I know?" and "What do I need to know?" will help them to decide what information is available.

For many problems which at first appear difficult, the solution becomes obvious once the information in the problem is grouped in an organized fashion. Experience with varied problem types suggests that how we organize information depends largely on the kind of information presented in the problem.

This unit examines three distinct, but closely related, methods of organization:

Lists — for problems which require enumerating multiple possibilities or combinations.

Tables — for problems in which one set of numbers bears a constant relationship to another. (ratios)

Charts — for problems which involve performing various operations (addition, subtraction, multiplication, division) based on inferences.

Often, organizing information into lists, tables, and charts leads to the discovery of *patterns* — regularly recurring relationships. This unit also focuses students' attention on recognizing and using patterns to seek solutions.

Often, too, the patterns which we observe can be summarized in a *rule* — a statement which describes the pattern and may be used to predict solutions to other similar problems. In this unit, students will have an opportunity to formulate and apply rules.

As students discuss the problems in this unit, encourage them to comment on how they solved each problem. They should also discuss why they chose one method of organization over another. They should explain what aspect of the problem led them to select a chart, or a list, or a table. This is an important step in helping them to analyze the *attributes* or conditions of the problem in answering the question "What strategy should I try?"

OBJECTIVES

Students will

- examine problems to select an appropriate method of organization.
- observe patterns and apply them in seeking solutions.
- formulate rules to predict solutions.
- solve problems by organizing complex information in lists, tables, or charts.
- create original problems which require organizing information.

OVERVIEW

Activity 1 Organizing information in exhaustive lists.
Activity 2 Discovering and elaborating on patterns.
Activity 3 Using tables to solve ratio problems.
Activity 4 Charting complex information to select operations.
Activity 5 Review of organizing strategies.

When students have completed this unit, you may wish to assign the Expert Problems, Set B, on pages 87-88, for cumulative strategy application.

Unit 3 • Strategy: Make a List, Table, or Chart

ACTIVITY 1

The Middle East
Student Book, pages 31 – 32

FOCUS: Using lists to enumerate all possible solutions
CONTENT CONNECTION: Regions (SOCIAL STUDIES)

All the problems in this activity require an exhaustive list — that is, a list of all the possible combinations. Some involve selecting any combination of two or three choices out of multiple possibilities. Others involve enumerating all the possible orders in which a set of objects can appear.

In every case, list-making will help students ensure that they have included all the possibilities. They will also be able to use their lists to observe patterns and formulate rules for solving similar problems.

Introduce list-making with Problem 1 by asking students what they need to know. (How many combinations of two topics there are.) Encourage them to estimate, predict, or even guess at the answer, and record their responses.

Then ask what some of the possible choices are. As students respond, indicate the choices on the board: BA, AE, HP, and so on. After several choices have been listed, ask, **How will we know when we have all the possibilities?** If no one suggests it, propose an *organized* or systematic list.

Group the various choices as follows:
```
BA
BE   AE
BH   AH   EH
BS   AS   ES   HS
BP   AP   EP   HP   SP
```

Make sure that students realize that AB and BA are not two different combinations, but merely two ways of indicating one combination. Once they have completed the list, the solution (15) should be obvious. You may want to compare the actual solution to the students' estimates.

One reason to construct the list as indicated is to help students observe a pattern and formulate a rule. Ask, **How many things did we have to choose among?** (six) **How many choices did each student have to select?** (two) Ask the class to look for a pattern. (They should see that each column has one less choice than the one to the left — 5, then 4, then 3, then 2, then 1, for a total of 15.)

Explain that another way to solve the problem is to start with a number that is one less than the number of items in the list — in this case 5, because there were 6 items — and add in order, the remaining numbers down to one.
(5 + 4 + 3 + 2 + 1 = 15)

Problem 2 may be solved in a similar way, by listing all the possibilities. There are only 10.

Problem 3 is similar, and really an extension of Problem 2. Grouping their lists, students should see a pattern of 10 + 6 + 3 + 1 = 20. This is, they will find, an extension of the 6 + 3 + 1 = 10 pattern of Problem 2. The difference is that Problem 2 involves choosing any 3 out of 5 and Problem 3 requires choosing any 3 out of 6. For choices involving 3 out of some number, the rule is that the column farthest right has 1 choice, the next column to the left, 3; then 6; then 10. The pattern is an increasing one: each column to the left is successively 2, then 3, then 4, then 5, then 6 larger than the one on the right. The total number of columns will always be 2 less than the total number of items.

Extension: Ask students to find how many possibilities there are for choosing any 3 items from a group of nine. The answer is 84:
(28 + 21 + 15 + 10 + 6 + 3 + 1 = 84)

In Problem 4, students will need to construct the full list and then eliminate the impossible combinations.

Problem 5 asks for the various sequences possible for 5 items. Using an organized list, students may see a pattern of 5 x 4 x 3 x 2 x 1 = 120. As an extension, ask them to guess how many ways 6 items could be arranged.
(720; 6 x 5 x 4 x 3 x 2 x 1 = 720)

CALCULATORS: *Students will find calculators helpful in applying the rules they discover.*

Before finishing this activity, be sure that students realize that organized lists are an excellent way of keeping track of all the possible arrangements, choices, or possibilities.

SOLUTIONS

1. 15 possibilities (5 + 4 + 3 + 2 + 1 = 15)

2. There would be fewer possibilities: only 10. If the possibilities are A, B, C, D, and E the combinations are:
   ABC   BCD   CDE   (6 + 3 + 1 = 10)
   ABD   BCE
   ABE   BDE
   ACD
   ACE
   ADE

3. 20 possibilities
   Rule: 10 + 6 + 3 + 1 = 20

4. 8 possibilities:
   SY EG IQ   EG IS SA
   SY EG IN   EG IQ SA
   SY EG SA   EG IN SA
   SY IQ SA
   SY IN SA

   The remaining 12 (of 20 total) possibilities are ruled out by the information in the problem.

5. 120 possibilities
   Rule: 5 x 4 x 3 x 2 x 1

ACTIVITY 1

NAME _____

The Middle East

1. In social studies, the class has been learning about ancient civilizations in the Middle East. Each team of students had to choose two civilizations as topics for a report. They could choose among the Babylonians, the Assyrians, the Egyptians, the Hittites, the Sumerians, and the Phoenicians. How many different combinations of two civilizations could they choose?

   THINK: How can you organize the information to be sure you have listed all the possibilities?

2. If there were only five civilizations to choose from, and each team had to select three, would there be more or fewer possible combinations? List each combination.

3. Last year, a trade mission visited six Mid-Eastern countries: Saudi Arabia, Yemen, Oman, United Arab Emirates, Qatar, and Bahrain. This year, the trade delegation can only visit 1/2 of those countries. What are their choices?

4. At the United Nations, Middle Eastern countries are trying to appoint a 3-member committee to plan peace talks. Candidates for the committee are Syria, Egypt, Israel, Iraq, Iran, and Saudi Arabia.

   If Israel is on the committee, neither Syria, nor Iraq, nor Iran will serve. If Iraq is a member, Iran will not be.

   How many possible combinations are there for the 3-member committee?

5. A senior diplomat needs to make 5 stops on a mission: Baghdad, Riyadh, Cairo, Damascus, and Amman. Leaving from Athens, how many different ways can he go if he makes only one stop in each city?

ACTIVITY 2

Addition Facts
Student Book, pages 33 – 34

FOCUS: Locating patterns in data
CONTENT CONNECTION: Number facts (MATHEMATICS)

This activity is an open-ended process of discovering as many patterns as possible in a set of data — here, addition facts to 18.

Allow students time to study the data. You may want to challenge teams to see which one can identify the most patterns. If students are slow to verbalize patterns, prompt them with questions.

In Problems 2-5, challenge students to test these patterns using other squares in the table. The patterns listed below are universal; that is, they are true for *any* square of the same size.

Extension: Students might want to try larger squares — 36, 49, or more — on their own to see what patterns emerge. After they have listed patterns for various size squares, suggest that they identify those patterns that are true for all squares, no matter what their size.

MANIPULATIVE: *You may wish to transfer the table to a transparency for the overhead projector so students can demonstrate patterns to the whole class.*

SOLUTIONS

1. Every row and column increases by one.

    Diagonals like this ⟋ consist of the same number; each increases by one.

    Diagonals like this ⟍ alternate between strings of consecutive odd numbers and consecutive even numbers.

CALCULATORS: *As many calculations will be necessary, you may want to let students use calculators to find patterns.*

2. The sums of the two diagonals are equal.
   The sum of the four numbers is twice the sum of a diagonal.
   The sum of the four numbers is four times the repeated diagonal number.
   The highest number in the square equals the lowest number plus two.

3. The sums of the diagonals are equal.
   The sum of each diagonal is three times the center digit.
   The sum of the nine digits equals the center digit multiplied by itself. (squared)
   The sum of the opposite corners is equal.
   The sum of the four corners is equal to the sum of the four remaining outside digits.
   The sum of the four corners equals four times the center digit.
   The sum of all nine digits equals three times the sum of each diagonal.
   The sum of the nine digits is the square of the center digit.

4. The sums of the diagonals are equal.
   The sums of the opposite corners are equal.
   The sum of the 16 digits is four times the sum of the diagonal.
   The sum of the outside digits is three times the sum of the inside digits.
   The sum of the outside digits is also three times the sum of a diagonal.
   The sum of the inside four digits equals the sum of a diagonal.
   The highest number is equal to the lowest number plus six.
   If you divide the square into four subsquares

   ```
   A | 5  6 | 7  8  | C
     | 6  7 | 8  9  |
     |------|-------|
   B | 7  8 | 9  10 | D
     | 8  9 | 10 11 |
   ```

   then the sum of the digits in subsquare A plus subsquare D equals the sum of the digits in subsquares B plus C.

5. The sum of each row or column is a multiple of five.
   The sums of the diagonals are equal.
   The sums of the opposite corners are equal.
   The highest number equals the lowest number plus eight.
   The sum of all 25 numbers equals twenty-five times the center number.
   The sum of the four corners equals the sum of the four middle outside numbers.
   The sum of all the outside numbers equals four times the sum of the corner numbers.
   The sum of the numbers in the inside square (9 numbers) equals nine times the center number.
   The sum of the outside numbers is 12 times the center number.

ACTIVITY 3

Into the Woods
Student Book, pages 35 – 36

FOCUS: Expressing ratios
CONTENT CONNECTION: Ecology (SCIENCE)

All the problems in this activity express ratios — a constant set of relationships between two sets of numbers. For this kind of problem, students will find tables a useful method of organization.

If necessary, prompt students to construct a table to record the information in Problem 1. They should quickly recognize the value of a table for the remaining problems.

CALCULATORS: *Filling in the details of the tables will be both quicker and more accurate with a calculator.*

Note: Some students may see these problems as essentially ones of simple operations. For example, in Problem 1 some may reason that 315 ÷ 45 = 7; 7 × 27 = 189. This is an entirely acceptable approach for students who understand the concept of ratios. In fact, this would be a good activity to use when the class is studying ratios.

In Problem 3, students will have to record not only the ratio but also the difference between the numbers of each kind of tree.

Problem 4 is not a ratio. The table simply serves to record data so that students can *Solve a Simpler Problem* — what times are sunrise and sunset four Fridays later.

In Problem 5, students will use a table to record both *new* and *existing* buds — since the solution depends on knowing both numbers.

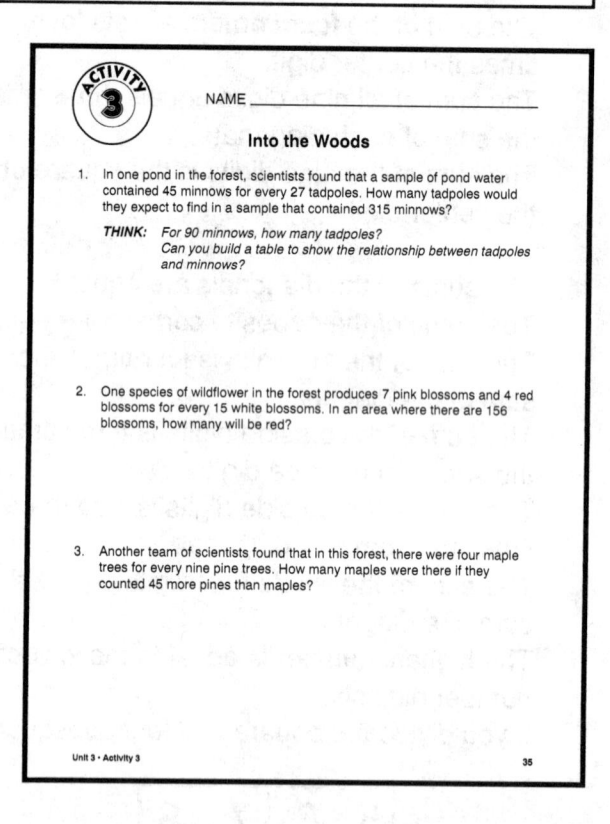

ACTIVITY 3

NAME _____

Into the Woods

1. In one pond in the forest, scientists found that a sample of pond water contained 45 minnows for every 27 tadpoles. How many tadpoles would they expect to find in a sample that contained 315 minnows?

    THINK: For 90 minnows, how many tadpoles?
    Can you build a table to show the relationship between tadpoles and minnows?

2. One species of wildflower in the forest produces 7 pink blossoms and 4 red blossoms for every 15 white blossoms. In an area where there are 156 blossoms, how many will be red?

3. Another team of scientists found that in this forest, there were four maple trees for every nine pine trees. How many maples were there if they counted 45 more pines than maples?

Unit 3 • Activity 3   35

SOLUTIONS

1. **189 tadpoles**

27	54	81	108	135	162	189
45	90	135	180	225	270	315

2. **24 red blossoms**

W	15	30	45	60	75	90
P	7	14	21	28	35	42
R	4	8	12	16	20	24

3. **36 maples**

P	9	18	27	36	45	54	63	72	81
M	4	8	12	16	20	24	28	32	36
dif	5	10	15	20	25	30	35	40	45

4. 14 hours, 12 minutes

Rise a.m.	6:25	6:18	6:11	6:04	5:57	5:50	5:43
Set p.m.	7:13	7:20	7:27	7:34	7:41	7:48	7:55

5:43 am to 7:55 pm = 14 hours, 12 minutes

5. 147 buds

Day	1	2	3	4	5	6	7
New	7	10	14	19	25	32	40
Total	7	17	31	50	75	107	147

---

4. As spring progresses, the length of the day steadily increases. One Friday, sunrise in the forest was at 6:25 AM, and sunset at 7:13 PM. The next Friday, the times were 6:18 AM and 7:20 PM. The following Friday, daylight dawned at 6:11 AM and the sun set at 7:27 PM. How many hours and minutes of daylight were there four Fridays later?

5. One day, a new plant had only 7 open leaf buds. The next day there were 10 more new open buds. The following day, 14 more new buds opened, and the next day, 19 more new ones opened. At the end of a week, if the pattern continued, how many leaf buds had opened on the plant?

ACTIVITY 4

Nature Walk
Student Book, pages 37 – 38

FOCUS: Using charts to organize complex information and selecting operations
CURRICULUM CONNECTION: Nature (ENVIRONMENT)

All the problems in this activity contain complex and unorganized information. Each problem, however, contains all the data necessary to fill in a chart, perform required operations, and reach a solution.

Start by having students read Problem 1. Many will be confused by the information. Suggest that they create on the chalkboard a large chart listing the types of trees and the various parks. Then as each piece of information from the problem is discussed, ask students to place it appropriately in the chart. Soon the chart will have the following data:

	Oak	Maple	Elm	Birch	T
Central	46	39			
Greenwood	24	13	32		85
Memorial				8	37
T	76		60		247

Now, students can begin to infer the operations necessary to fill in the chart. For example, 247 – 37 – 85 = 125 trees in Central Park. Also 76 – 24 – 46 = 6 oaks in Memorial Park, and therefore 12 maples, leaving 11 elms. In a similar way, students can complete this table and construct tables for the remaining problems of this activity.

As students work with these problems, reassure them that they are not difficult simply because they are long. The key is to set up a chart so that information can be recorded as it occurs.

Note: Some of the problems will include zeros in the tables; it is important to record them so that totals will add up and correct inferences can be made.

CALCULATORS: Students will find calculators very helpful in filling in their tables.

ACTIVITY 4

NAME _____

Nature Walk

1. The class is doing a census of trees in the town parks. In Central Park, they counted 46 oaks, 39 maples, some elms and some birches. In Greenwood Park, they found 32 elms, 3/4 as many oaks as elms, some birches, and 1/3 as many maples as in Central Park, for a total of 85 trees.

    Memorial Park is the smallest park in town, with only 37 trees. It has 8 birches, some elms, and 1/2 as many oaks as maples. In the three parks, there are 60 elm trees in all, 16 more oaks than elms, and a total of 247 trees of all four kinds. How many elms are there in Memorial Park?

    THINK: How many kinds of trees are there? How many parks? Would a chart help us figure out how many trees of each kind are in each park?

2. Memorial Park is noted for its four types of flowering trees. Some are pink, and some are white. There are 180 flowering trees in all, and 3/5 of them are pink. All the crabapple and cherry trees have pink flowers. There are 27 crabapple trees. Some of the dogwood trees are pink, and some are white. There are 70 of them altogether, and 2/5 of them are pink. All the apple trees are white. How many more cherry trees than apple trees are there?

Unit 3 • Activity 4

SOLUTIONS

1. 11 elm trees

	Oak	Maple	Elm	Birch	Totals
Central	46	39	17	23	125
Greenwood	24	13	32	16	85
Memorial	6	12	11	8	37
Totals	76	64	60	47	247

2. 23 more cherry trees

	Pink	White	Totals
Crabapple	27	0	27
Cherry	53	0	53
Dogwood	28	42	70
Apple	0	30	30
Totals	108	72	180

3. 90 red flowers

	white	red	pink	purple	yellow	Totals
Tulip	108	72	108	36	36	360
Iris	12	18	36	78	36	180
Daffodil	108	0	0	0	12	120
Crocus	12	0	36	6	6	60
Totals	240	90	180	120	90	720

4. more purple irises

5. 12 white crocus, no red, 36 pink, 6 purple, 6 yellow

Workers are planting bulbs in one of the beds in Central Park. They used a total of 720 bulbs of four kinds: crocus, tulip, daffodil, and iris. The flowers are of 5 different colors: white, red, pink, purple, and yellow. One-half of the bulbs are tulips; one-third of the bulbs are white. One-fourth of the bulbs are irises; one-fourth are pink. One-sixth of the bulbs are purple; one-sixth are daffodils. There are the same number of red and yellow bulbs.

3. How many red flowers are there in all?

One-fifth of the tulips are red. There are exactly as many purple tulips as yellow ones. Together, there are as many purple and yellow tulips as red ones. There are as many yellow irises as yellow tulips, and as many pink irises as yellow ones. There are half as many red irises as pink ones, and two-thirds as many white irises as red ones.

4. Are there more purple tulips or purple irises?

There are as many pink tulips as white daffodils. There are only two colors of daffodils, white and yellow. Twelve of the daffodils are yellow.

5. How many of each color crocus bulb are there?

ACTIVITY 5

Challenge
Student Book, pages 39 – 40

FOCUS: Review

This activity reviews the various organizing strategies taught in the unit. These problems can serve as an assessment of students' understanding. They could also be assigned to small groups of students as additional practice. You may want to assign tasks to each group member to provide a cooperative learning group setting.

In discussing solutions, ask students to explain why one way of organizing data seemed to work for each problem.

SOLUTIONS

1. The digits in each product add up to 9.
   (1 + 8, 2 + 7, 3 + 6, etc.)
   In a list of the products, the first digit increases by one and the second digit decreases by one each time: 18, 27, 36, 45, etc.

2.

	Cole Slaw	Potato Salad	Fries	Chips	Totals
Ham	24	51	84	32	191
Tuna	70	38	90	30	228
Cheese	84	132	192	28	436
PJ	36	45	40	244	365
Totals	214	266	406	334	1220

3. Green   36
   Black   144
   White   96
   Blue   32
   Red   16

4. 120 possibilities (Pattern: 5 x 4 x 3 x 2 x 1 = 120)

5. Football (85 cards)

	Beto	Phil	Rayna	Totals
Baseball	37	30	15	82
Basketball	0	20	41	61
Hockey	26	13	13	52
Football	17	0	68	85
Totals	80	63	137	280

ACTIVITY 5

NAME _____

Challenge

1. Multiplication tables are a type of pattern. Think about the "9-times table." How is it different from others? What special patterns does it contain?

2. The cafeteria has a Sandwich Special, which is any combination of a sandwich and a side dish. There are four kinds of sandwiches: tuna, cheese, ham, and peanut butter/jelly. There are four side dishes: chips, fries, cole slaw, or potato salad. Last week 1,220 Sandwich Specials were sold. Peanut butter/jelly and chips was the most popular combination, making up 1/5 of all the specials sold. Cheese was the most popular sandwich; 436 were sold. Fries — of which 406 were sold — was the most popular side dish. The least popular combination was ham and cole slaw — only 24 were sold.

   They sold 132 cheese/potato salad combinations, and 60 more than that of cheese sandwich and fries. Of the remaining cheese sandwiches, 3/4 were served with cole slaw, the rest with chips. Chips were served with ham 32 times. In all, 334 servings of chips were served. For every serving of tuna and chips, they sold 3 servings of tuna and fries. Peanut butter/jelly was served with fries 40 times, 5 servings less than with potato salad. Peanut butter/jelly specials were served a total of 365 times, and cole slaw 214 times. If tuna was served with potato salad 38 times, how many of each combination were sold?

3. Last week Monarch Motor Car factory produced 324 cars. One-ninth of all the cars were green. For every green car they built, they turned out 4 black ones, and for every 3 black ones, 2 white ones. Of the remaining cars, 2/3 were blue and the rest were red. How many of each color did they produce?

4. You are doing a report on five Latin American countries: Mexico, Chile, Bolivia, Guatemala, and Brazil. One section of your report is on each country. How many different ways could you arrange the order of the sections?

5. Beto, Phil, and Rayna collect sports cards. Altogether they have 280 cards. Some are basketball players, some are hockey players, some are football players, and some are baseball players. Beto does not collect basketball cards, but has 37 baseball cards. He also has some football cards and half of all the hockey cards in the group. There are a total of 52 hockey cards.

   Phil has 3 kinds of cards, including 2/3 as many basketball cards as baseball cards. He has 63 cards in all; 30 are baseball cards. Rayna has as many hockey cards and half as many baseball cards as Phil, along with 41 basketball cards and 27 more football cards than basketball cards.

   In all, they have 61 basketball cards and 85 football cards. What kind of cards do they have the most of?

Strategy Discussion

Ask students to explain the characteristics of problems in which each kind of organizing method was helpful.

Lists — to count up all the possible combinations.
Tables — to compare one number to another in a pattern.
Charts — to make sense out of complicated information and decide what numbers to add or subtract, multiply or divide.

Conclude the discussion by asking, **What is similar about all three of these methods?** (They are all ways of organizing information. They make it easier to use or remember information.)

Extension

Students may enjoy creating their own "puzzle book" to share with classmates.

These are challenging problems to create, since too much information can make the problem meaningless, while too little information in clues makes them impossible to solve.

Making Lists, Tables, or Charts
Student Book, pages 41 – 42

This page of the student text is designed to be taken home as a family involvement activity.

Included for the parent is the brief explanation of the unit content with a summary of the strategy which has been developed.

Along with that background information, there are several problems which the student and the family can solve together.

This page may be assigned as soon as Activity 4 of this unit has been completed.

SOLUTIONS

1. 15 combinations (5 + 4 + 3 + 2 + 1)

```
VAN CHOC
VAN ST    CHOC ST
VAN PCH   CHOC PCH   ST PCH
VAN MNT   CHOC MNT   ST MNT   PCH MNT
VAN BAN   CHOC BAN   ST BAN   PCH BAN   MNT BAN
```

2. 56 raisins

raisins	8	16	24	32	40	48	56
peanuts	13	26	39	52	65	78	91

3. 51 green square blocks

	square	rectangle	triangle	Totals
blue	4	80	112	196
red	57	27	14	98
green	51	26	21	98
Totals	112	133	147	392

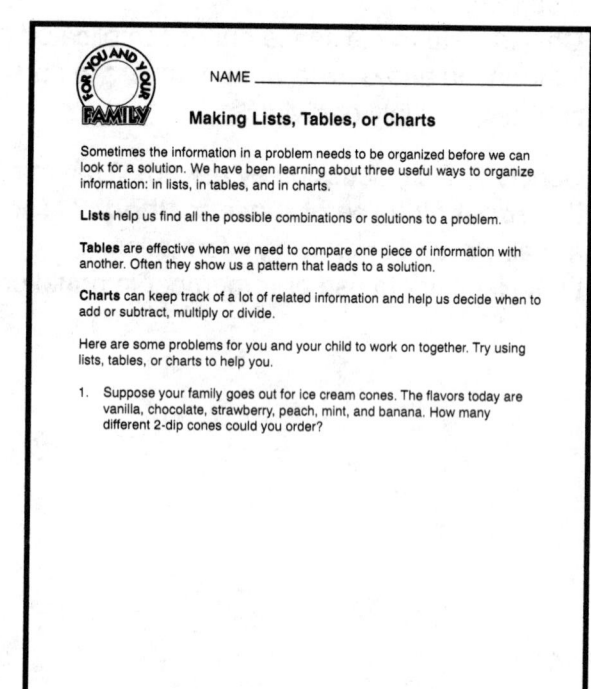

Making Lists, Tables, or Charts

Sometimes the information in a problem needs to be organized before we can look for a solution. We have been learning about three useful ways to organize information: in lists, in tables, and in charts.

Lists help us find all the possible combinations or solutions to a problem.

Tables are effective when we need to compare one piece of information with another. Often they show us a pattern that leads to a solution.

Charts can keep track of a lot of related information and help us decide when to add or subtract, multiply or divide.

Here are some problems for you and your child to work on together. Try using lists, tables, or charts to help you.

1. Suppose your family goes out for ice cream cones. The flavors today are vanilla, chocolate, strawberry, peach, mint, and banana. How many different 2-dip cones could you order?

2. The Sweetheart Candy Factory makes a mix of chocolate-covered peanuts and chocolate-covered raisins. They use 8 raisins for every 13 peanuts. If there are 91 peanuts in the box of candy, how many raisins are there?

3. Little Joe has 392 blocks. Of these, 2/7 are square, 1/4 are red, and 3/8 are triangles. Of the blocks that are not red, 2/3 are blue. Only 4 of the blocks are blue squares. Of the remaining blue blocks, 5/12 are rectangles. Only 1/7 of the triangles are green. If there are 27 red rectangles, how many green square blocks does he have?

Unit 3 • For You and Your Family

UNIT 4

STRATEGY: Guess and Check

INTRODUCTION

For certain types of problems, the most effective strategy is to make an assumption — a *guess* — then test it — *check* — against the conditions in the problem. The first guess is often wrong, but it can be used to make a reasonable second guess which can bring you closer to the solution.

This *Guess and Check* strategy is particularly appropriate for problems which require identifying numbers that fit an arbitrary set of conditions. *Guess and Check* provides a starting point for testing possible solutions for problems that require finding

- numbers which total a given sum and product.
- numbers with stated attributes such as greatest common factors or least common multiples.
- combinations of groupings which produce a given total.
- a series whose component members can be identified.

As they solve the problems in this unit, students will realize that their guessing should be *persistent;* that is, they should use the results from one guess to make successive guesses, adjusting their guesses by the results they obtain.

OBJECTIVES

Students will

- make a series of reasonable guesses in search of a solution.
- adjust guesses on the basis of results obtained.
- persist in guessing until a solution is reached.
- identify elements in a problem which make *Guess and Check* an appropriate strategy.
- create original *Guess and Check* problems.

OVERVIEW

Activity 1 Systematic guessing to produce a given total.
Activity 2 Systematic guessing involving complex information.
Activity 3 Guessing to determine combinations totaling a given number.
Activity 4 Guessing numbers which meet given operational conditions.
Activity 5 Review of problem types.

When students have completed this unit, you may wish to assign the Expert Problems, Set C, on pages 89-90, for cumulative strategy application.

ACTIVITY 1

On the Court
Student Book, pages 45 – 46

FOCUS: Guessing to determine ratios
CONTENT CONNECTION: Basketball (PHYSICAL EDUCATION)

The first problem of this activity should be fairly easy for most students. It is included here to underscore the value of persistent and systematic guessing, a strategy students will later apply to more complex problems.

Students need to realize that they must first *Solve a Simpler Problem* — What is the difference between the two scores?

Since a basket is worth two points and the difference between scores was four baskets, there was an eight point difference.

Remind students that we now know two important facts: the sum of the two numbers (230) and the difference between them (8). Ask, **What should we do now to find the two numbers?** (Make a guess.) **Would 50 and 300 be a good guess?** (no — too large)

Ask volunteers to make a guess, one at a time. Record each guess on the board and check it with the information in the problem. **What two things must each pair of numbers do?** (add to 230; have a difference of 8)

As each guess is checked, decide if the next guess should be larger or smaller. Proceed until the correct solution (119 to 111) is identified.

Problems 2-5 are similar, in that there is a known total. The task in each, however, is to identify the mix of elements which constitute that total. In Problem 2, for example, some of his scores were worth 1 point, some worth 2 points, and some worth 3 points.

To make sure that students understand this concept, you may want to develop the solution with the whole class. As a starting point, ask, **What do we need to know?** (how many of each kind of score he made)

What kinds of scores did he make? (foul shots, baskets, 3-pointers) **How many points is each worth?** (1, 2, and 3 points)

What do we already know? (He made 2/3 as many foul shots as baskets.) **What else?** (4 times as many points for baskets as for 3-pointers)

As students begin guessing, ask, **Does anyone have an idea how to keep track of our guesses?** If no one suggests the *Make a List or Table* strategy, introduce this format:

```
FOUL     1   x _____ = _____
BASKET   2   x _____ = _____
3-PT     3   x _____ = _____
                        TOTAL
```

Have students check each guess. Does it add up to the correct total? Are there 2/3 as many foul shots as baskets? Are there 4 times as many basket points as 3-pointers?

Again, persistent guessing will produce a solution, as it will in the remaining problems of the activity.

CALCULATORS: *Throughout this unit, the use of calculators will produce faster and more accurate results, thus sustaining student perseverance.*

SOLUTIONS

1. Score: 119 to 111

2. He made 4 foul shots.

    6 baskets   = 12 points (4 times as many as for 3-pointers)
    4 foul shots = 4 points (2/3 of 6)
    1 3-pointer = 3 points
                    19 points

3. 11 baskets   = 22 points
    5 foul shots = 5 points
    2 3-pointers = 6 points
    18 scores   = 33 points

    or

    7 baskets   = 14 points
    4 3-pointers = 12 points
    7 foul shots = 7 points
    18 scores   = 33 points

4. He made 12 three-pointers.

    12 3-pointers × 3 = 36
    126 baskets   × 2 = 252
    138   288

5. They sold 6,563 regular tickets and 3,853 discount tickets.

    Regular at $22 = 6,563 = $144,386
    Discount at $17 = 3,853 = $ 65,501

Unit 4 • Activity 1

Energy Sources
Student Book, pages 47 – 48

FOCUS: Guessing with known sums and differences
CONTENT CONNECTION: Energy (SCIENCE)

This activity extends students' experiences with *Guess and Check* problems of the kind found in the previous activity. Here, they will need to distinguish between problems which provide clues of sums and differences (Problems 3 and 4) and those which include information as to the mix of elements which add up to a given total.

All of the problems are solvable with the *Guess and Check* strategy developed in the previous activity.

Note: Two of the problems use technical units of measurement. The problems have been constructed so that students do not have to use the full numbers, but can combine smaller numbers with the abbreviations MST (for million short tons) and BB (for billion barrels) in seeking their solutions.

In Problem 3, students will need to discover, then *Solve a Simpler Problem*: How much coal was used to generate electricity in 1965? They can then *Work Backwards* to a solution.

ACTIVITY 2 — Energy Sources

1. Tonio works Saturdays at the local service station. He spent most of the morning filling small containers with gas for lawn mowers. Some of the containers held 2 gallons and some held 3 gallons. He filled 38 tanks with a total of 87 gallons. How many 3-gallon tanks did he fill?

    THINK: What combination of 2-gallons and 3-gallons will fit both conditions — 38 containers in all, with a total of 87 gallons?

2. In one recent year, Americans spent $95 billion on energy sources (oil, gas, and electricity) to heat and cool their homes, operate appliances, and cook their meals. They spent $36 billion more on electricity than on gas. For the two fuels together, they spent $88 billion. How much more did they spend on gas than on oil?

3. Coal consumption has increased in the last quarter-century. In 1965, Americans used 270 MST (million short tons) of coal to heat buildings and produce electricity. Of that, 220 more MST were used to generate electricity than to heat buildings. Only 25 years later, they used 19 MST less to heat than in 1965, but 526 more MST to produce electricity. How much more coal was used in 1990 for heat and electricity than in 1965?

SOLUTIONS

1. 11 three-gallon tanks

    ```
      27  2-gal =    54
    + 11  3-gal =  + 33
    ----           ----
      38             87
    ```

2. $19 billion more

gas	$26 billion	gas	$26 billion
elec	$62 billion	oil	− $7 billion
		difference	$19 billion

3. 507 MST more

	1990	1965
electricity	771	245
heat	+ 6	+ 25
	777	− 270 = 507

4. 30 BB more

   USA, 27; Mexico, 52; China, 22; Russia, 58; Venezuela, 60

5. 210 cars and 54 bicycles

   cars   210   x   4 wheels   =   840
   bicycles + 54 x   2 wheels   =   +108
             264   948

---

4. Although most of the world's oil reserves are in the Middle East, other countries have reserves of their own. For example, Mexico and the United States together have a reserve of 79 billion barrels (BB). Mexico has 25 BB more than the United States. Together, China, Russia, and Venezuela have another 140 BB. Russia has 36 BB more than China, while Venezuela has 38 BB more than China. How much more oil does Mexico have than China?

5. Using mass transit helps to save energy. Riding a bicycle instead of driving helps to save on gasoline and so does riding a train. In the commuter parking lot at the local railroad station, there were 264 vehicles one morning. Altogether there were 948 wheels. How many of the vehicles were cars, and how many were bicycles?

ACTIVITY 3

Cathedral
Student Book, pages 49 – 50

FOCUS: Guessing a series which produces a known total
CONTENT CONNECTION: Middle Ages (SOCIAL STUDIES)

This activity further reviews and extends problems for which a *Guess and Check* strategy is appropriate.

Problem 1 requires identifying two numbers whose product is given. Also given is the difference between them. Help students to recognize that this problem is similar to one in which the sum and the difference of the numbers are given. Students might begin by finding pairs of numbers with a difference of 16, then finding their product.

Problems 2 and 3 introduce a new type of *Guess and Check* problem — one in which a series of numbers adds up to a given total. Also known is the number of items in the series, and the difference between each number in the series and the next number.

Help students begin Problem 2 by asking the questions, **What do we need to know?** (how many stones each day) **What do we already know?** (total number of stones; number of days) **What else?** (15 more stones each day) A good strategy is to first guess the number of stones for the middle day. Students can then add or subtract to get the total for the other days.

Solutions for problems of this type are not difficult, although they require multiple guesses.

After students have completed Problem 2, show them a simple rule for solving problems of this type. Divide the total (550) by the number of items which make up the total (in this case, 5, the number of days). The result (110) is the answer for the middle term in the series — that is, the third of the five days. It is then a simple matter to add and subtract to get the answers for the remaining terms.

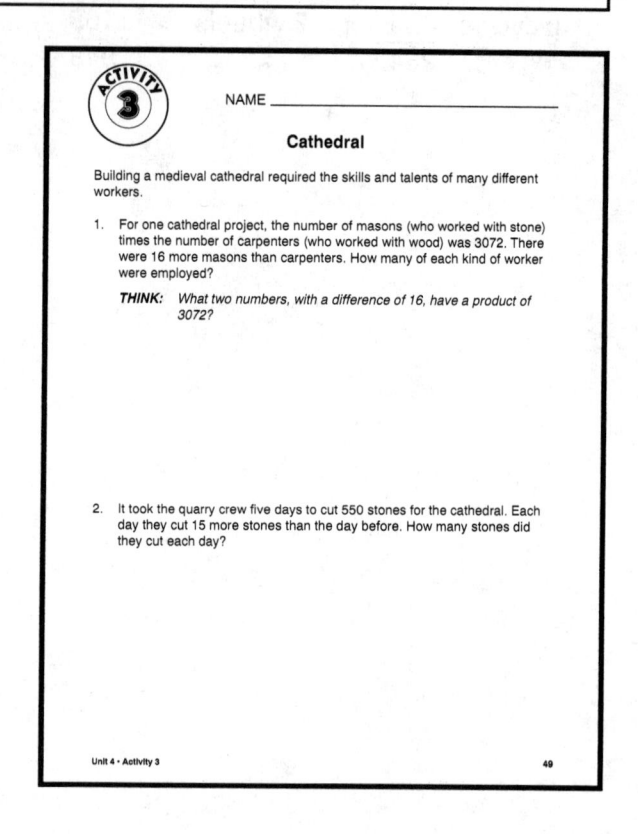

ACTIVITY 3

NAME _____

Cathedral

Building a medieval cathedral required the skills and talents of many different workers.

1. For one cathedral project, the number of masons (who worked with stone) times the number of carpenters (who worked with wood) was 3072. There were 16 more masons than carpenters. How many of each kind of worker were employed?

   THINK: What two numbers, with a difference of 16, have a product of 3072?

2. It took the quarry crew five days to cut 550 stones for the cathedral. Each day they cut 15 more stones than the day before. How many stones did they cut each day?

Unit 4 • Activity 3   49

Let students try this approach for Problem 3. Problems 4 and 5 are more complex applications of finding the mix, or proportions, which make up a total.

CALCULATORS: *All of these problems will be easier to solve if students use calculators to check their guesses.*

SOLUTIONS

1. 64 masons; 48 carpenters

2. Day   1   2   3   4   5
         80 + 95 + 110 + 125 + 140 = 550

3.

Day
  1   2   3   4   5   6   7   8   9
120 + 150 + 180 + 210 + 240 + 270 + 300 + 330 + 360 = 2,160

4.
```
    red      27
 yellow      81
   blue   + 324
           432
```

5.
```
    6 doors of walnut  (64 shillings)  =   384
  + 3 doors of ash     (85 shillings)  = + 255
    9                                      639
```

Unit 4 • Activity 3   45

ACTIVITY 4

Math Team
Student Book, pages 51 – 52

FOCUS: Inferring numbers from operational clues
CONTENT CONNECTION: Number Facts (MATHEMATICS)

This activity introduces several variations on *Guess and Check* problems. Persistent guessing will provide solutions to all of them.

Problems 1 and 2 require students to guess the numbers to insert into partially completed problems.

To help students get started, ask them to read Problem 1. Ask them if there are any clues to help them. Direct their attention to the last number in the solution. Ask what number in the ones' place in the multiplier will produce an 8 here. (only a 2) Give them time to try out various guesses.

Problem 2 is similar, except that the operation involved is division.

Problems 3 and 4 depend on understanding two concepts:

Least common multiple (LCM) — the smallest number that is a multiple of a set of numbers. See the following example:

Multiples of 3	Multiples of 4	Multiples of 6
3	4	6
6	8	12
9	12	18
12	16	(24)
15	20	
21	(24)	
(24)		

Least common multiple

Greatest common factor (GCF) — the largest number that divides evenly into two given numbers.

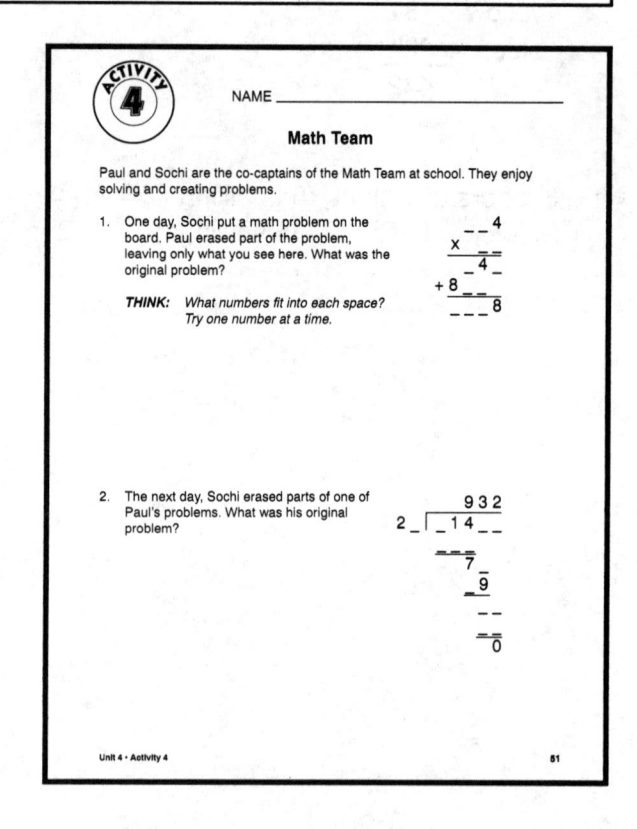

Once students understand these terms, these problems are similar to those in which sums, differences, and products are known.

Problem 5 requires only two things: knowledge of addition facts and persistent guessing.

SOLUTIONS

1.

```
    274        or      424        or      224
  x  32              x  22              x  42
   548                848                448
 + 822              + 848              + 896
  8768               9328               9408
```

46   Unit 4 • Activity 4

2.

```
        932
    ┌────────
23 │ 21436
     207
     ───
      73
      69
      ──
       46
       46
       ──
        0
```

3. Two possible answers are: 4 and 24 or 8 and 12. (If students find only the obvious choice, 4 and 24, challenge them to find another solution.)

4. The numbers are 10 and 15.

5.

				50
12	5	1	16	34
6	15	11	2	34
3	10	14	7	34
13	4	8	9	34
34	34	34	34	50

---

3. Paul wrote down two numbers. He told Sochi that their least common multiple (LCM) is 24 and their greatest common factor is 4. What were his two numbers?

4. Sochi also wrote down two numbers. She told Paul that their greatest common factor was 5, and their least common multiple was 30. She also said that one of the two numbers was 2/3 of the other. What were the two numbers?

5. Sochi and Paul made up a puzzle. The numbers from 1-16 are arranged in the puzzle so that the sum of each row and each column is the same. Also, the two diagonals have the same sum, although this is a different number than the sum of the rows and columns. Paul and Sochi have arranged some of the numbers for you. Complete the puzzle.

	5		16
6			
		14	
13			9

Unit 4 • Activity 4

ACTIVITY 5

Challenge
Student Book, pages 53 – 54

FOCUS: Review

These problems can serve as an assessment of students' understanding. They could also be assigned to small groups of students as additional practice. You may want to assign tasks to each group member to provide a cooperative learning group setting.

SOLUTIONS

1.

Grilled cheese	$1.49		$5.00
Fruit cup	$1.19		− $3.83
Milk	+ $1.15		$1.17
	$3.83		

or

Hot dog	$1.89
Chips	$ .79
Milk	+ $1.15
	$3.83

or

Tuna	$1.69
Chips	$ .79
Soft Drink	+ $1.35
	$3.83

2. 53 and 97

Subtract the difference from the sum and divide by two. The result is the smaller number. Add the difference to get the larger number.

```
  150        106 ÷ 2 = 53        53
− 44                            + 44
  106                             97
```

3. a. 17 and 86
   b. 22 and 47

4. Day 1   2   3   4   5   6   7
       14 + 36 + 58 + 80 + 102 + 124 + 146 = 560

5. 18 and 24

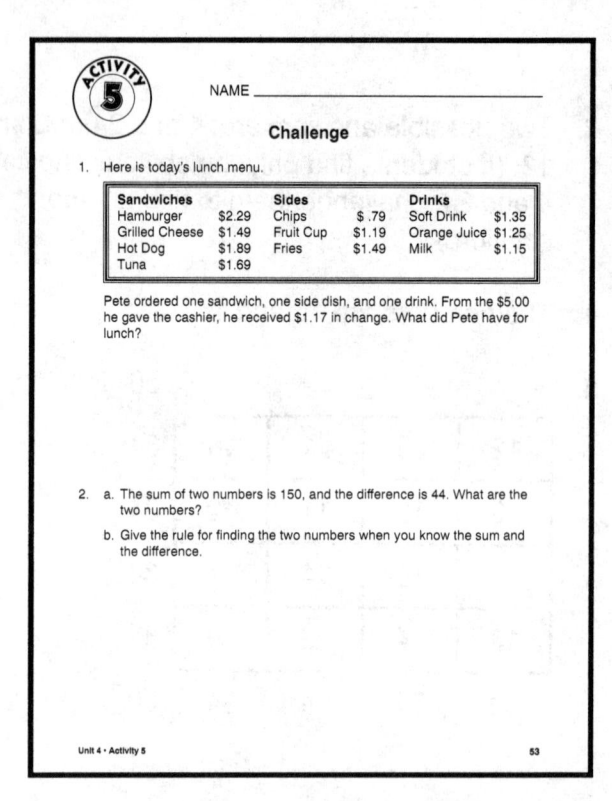

ACTIVITY 5

NAME _____

Challenge

1. Here is today's lunch menu.

Sandwiches		Sides		Drinks	
Hamburger	$2.29	Chips	$ .79	Soft Drink	$1.35
Grilled Cheese	$1.49	Fruit Cup	$1.19	Orange Juice	$1.25
Hot Dog	$1.89	Fries	$1.49	Milk	$1.15
Tuna	$1.69				

Pete ordered one sandwich, one side dish, and one drink. From the $5.00 he gave the cashier, he received $1.17 in change. What did Pete have for lunch?

2. a. The sum of two numbers is 150, and the difference is 44. What are the two numbers?
   b. Give the rule for finding the two numbers when you know the sum and the difference.

3. a. The sum of two numbers is 103. Their product is 1,462. What are the two numbers?
   b. The product of two numbers is 1,034. If the difference between them is 25, what are the two numbers?

4. It took the factory an entire week to produce 560 of the new model cars. Each day, they turned out 22 cars more than the day before. How many cars did they manufacture each day?

5. The greatest common factor of two numbers is 6. The least common multiple of those two numbers is 72. The difference between the two numbers is 1/3 of one of the numbers and 1/4 of the other number. What are the two numbers?

STRATEGY DISCUSSION

Ask students why making a guess was helpful in solving the problems in this unit. (Possible response: It provides a place to start.) **How did you decide what numbers to try?** (estimated to make a reasonable guess) **What did you do next?** (checked to see if the guess fit the problem)

How did wrong answers help you? (The answer could be used to decide whether to guess a larger or smaller number the next time.) **Did it take several guesses?** (usually, yes)

Ask students to recall the various kinds of problems which they solved in this unit.

EXTENSION

Some students may enjoy making up original problems like the ones in this unit. You may want to review concepts such as *least common multiple* and *greatest common factor* and challenge students to create problems of their

Guess and Check
Student Book, pages 55 – 56

This page of the student text is designed to be taken home as a family involvement activity.

Included for the parent is a brief explanation of the unit content with a summary of the strategy which has been developed.

Along with that background information, there are several problems which the student and the family can solve together.

This page may be assigned as soon as Activity 4 of the unit has been completed.

SOLUTIONS

1. Chicken wings $ 4.25
    Fried fish $13.00
    Pie   + $ 2.75
                          $20.00   x   15%   = $3.00

2. Score: 123-116

3. on page 86

4. 64 and 95

5. 265 bricks

    Day   1   2   3   4   5
            145 + 175 + 205 + 235 + 265 = 1,025

Guess and Check

Sometimes the best way to solve a problem is to make a good guess and see if it works!

We have learned that it helps to make a *reasonable* guess — one that we think might be a possible answer. We have also learned that even a wrong guess can be useful because it can help us decide whether our next guess should be larger or smaller.

Finally, we have learned to keep guessing until we reach a solution.

Here are a few problems, like the ones we have been solving, for you to try at home.

1. This is a menu from a local restaurant.

Appetizer		Main Course		Dessert	
Soup	$3.00	Steak	$18.50	Pie	$2.75
Shrimp Cocktail	$5.75	Pork Chops	$15.50	Cake	$3.25
Chicken Wings	$4.25	Fried Fish	$13.00	Ice Cream	$2.25

    Suppose you ordered an appetizer, a main course, and a dessert. When your check comes, you leave a $3.00 tip (exactly 15%). What did you have for dinner?

2. In the finals of the basketball tournament, 239 points were scored in all. The Jaguars beat the Barons by 7 points. What was the score?

3. A magazine article covered five consecutive pages. If the sum of the pages is 440, on what page did the article start?

4. Two numbers have a product of 6,080. The difference between them is 31. What are the two numbers?

5. A new bricklayer is learning the trade. In 5 days, he laid 1,025 bricks. Each day, he laid 30 bricks more than the day before. How many bricks did he lay on the fifth day?

Unit 4 • For You and Your Family

UNIT 5

STRATEGY: Use Logical Reasoning

INTRODUCTION

Logical reasoning, as presented in this unit, is the process of organizing and analyzing information. Students will be encouraged to make reasonable inferences, draw accurate conclusions, and reach defensible solutions.

In these problems, students will need to use directly stated information as clues to additional, though not directly stated, information. Once all the facts, stated and unstated, have been determined, solutions which contradict those facts can be eliminated. The correct solution should then become obvious.

These types of problems may pose difficulties for some students. For example, students may lack the confidence to read carefully through the whole problem. Others may be unsure of how to draw logical conclusions from what they read. Still others may not recognize the value of negative information — that knowing what is not true may help us discover what *is* true.

All of these problems can be addressed by having students work through the problem one sentence at a time, stopping after each phrase to discover what the sentence means and implies, and to decide what is known and what can be determined from the known information.

As students gain confidence and experience, they will work more independently. For every problem, however, they should be prepared to discuss how they arrived at their solutions. Such discussion helps them to acquire, broaden, and internalize patterns of logical reasoning.

OBJECTIVES

Students will:
- locate relevant information in a problem.
- use clues presented in the problem to infer additional information.
- develop appropriate lists, tables, or charts to keep track of information, both stated and inferred.
- support solutions with logical reasoning.

OVERVIEW

Activity 1 Locating clues and organizing information.
Activity 2 Using negative information; constructing truth tables.
Activity 3 Constructing tables.
Activity 4 Logical reasoning.
Activity 5 Review of various problem types.

When students have completed this unit, you may wish to assign the Expert Problems, Set D, on pages 91-92, for cumulative strategy application.

ACTIVITY 1

Playoff Season
Student Book, pages 59 – 60

FOCUS: Making inferences; organizing information
CONTENT CONNECTION: Sports contests (PHYSICAL EDUCATION)

This activity is designed to encourage students to look for clues as a means of making inferences. The first problem is simply addition and subtraction, once students recognize the relationship among the numbers. The value of this problem lies in the opportunity for students to explain how they decided to arrange the numbers. Most students will be able to solve this problem without prompting, though not all will be able to verbalize the process. Help them see the logic of the statement "If the Tigers, who won 12 games, won 3 more than the Cardinals, then the Cardinals won 3 games less than the Tigers, or 9 games."

Problem 2 is similar, but involves the use of fractions.

Problem 3 presents a visual organizer to help students analyze the information. Explain that the winners of each round advance to the next round. Once a team loses a game, it is eliminated from the tournament. Therefore, teams who won no games do not advance. With this information, students should be able to fill in the chart for Round 2. Ask, **How many games did the champion win?** (3 games) **What is 1/3 of 3?** (one) **Who won only one game?** (Panthers and Tigers) **Who won two games?** (Lions) **How do you know?** (The others won half as many as the Lions.) **Who is left as the champion?** (Orioles)

Problem 4 demonstrates the value of negative information. Ask, **If Megan had Fred's bat, whose glove did she have?** (Cooper's. She did not have her own and she did not have Fred's, since no one had two items belonging to the same person.) A chart might help students visualize the solution.

	Glove	Bat
Megan	Cooper's	Fred's
Fred	Megan's	Cooper's
Cooper	Fred's	Megan's

Fred must have Megan's glove since he does not have his own or Cooper's. Therefore, he has Cooper's bat. That leaves only Fred's glove for Cooper to have along with Megan's bat.

Problem 5 represents another type of clue — "mutual exclusivity." If we know what is eliminated, we can figure out what is left.

MANIPULATIVES: *This is an excellent type of problem to demonstrate with a manipulative. Use three paper bags, the appropriate labels, and objects to represent the apples and oranges.*

The strategy for all problems of this type is to take one object from the bag mismarked as containing the mixture. In this case, the bag is wrongly labeled *Apples and Oranges*. The logic is this:

If she took an apple from the bag labeled *Apples and Oranges*, we know the labels are wrong so there are really just apples in that bag. The label on the bag labeled *Apples* cannot have apples in it so it must have oranges in it, and the bag labeled *Oranges* must have the apples and oranges.

	A/O	A	O	
				(wrong labels)
If A removed, then	A	O	A/O	(correct labels)
If O removed, then	O	A/O	A	(correct labels)

With all the problems in this activity, be sure to ask students to explain in detail how they arrived at their solutions. It is important for them to verbalize their thinking processes.

SOLUTIONS

1. The Orioles won 15 games.

    12 − 3 = 9
    9 + 12 = 21
    36 − 21 = 15

2. The Cougars scored 19 more runs than the Broncos.

    114 ÷ 2 = 57   Lions
    57 ÷ 3 = 19   Broncos
    57 + 19 = 76
    114 − 76 = 38   Cougars
    38 − 19 = 19

3. The Orioles, who won 3 games, won the tournament.

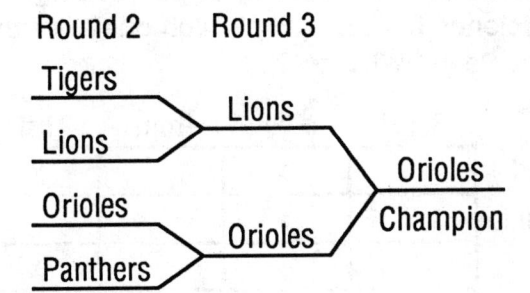

    Round 2   Round 3
    Tigers
         ⟩ Lions
    Lions
                    ⟩ Orioles
    Orioles   Champion
         ⟩ Orioles
    Panthers

4. Cooper had Fred's glove.

5. She took one piece of fruit from the bag mismarked *Apples and Oranges*.

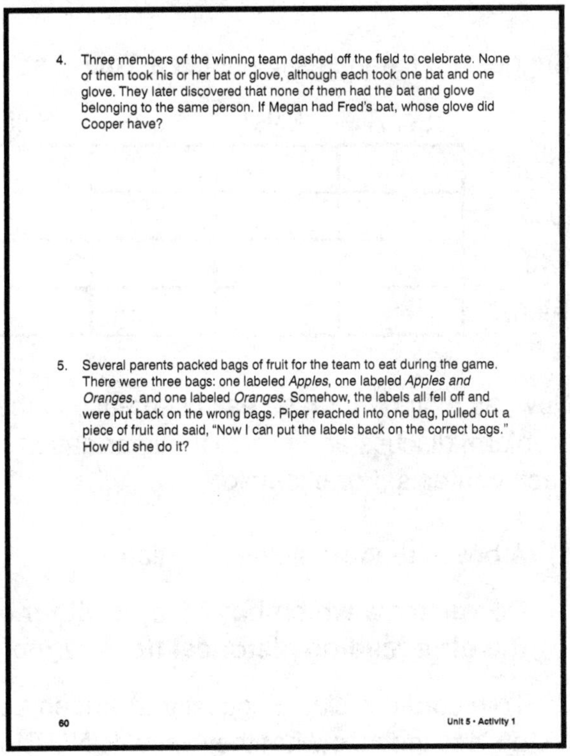

	A/O	A	O	(wrong labels)
If apple, then	A	O	A/O	(correct labels)
If orange, then	O	A/O	A	(correct labels)

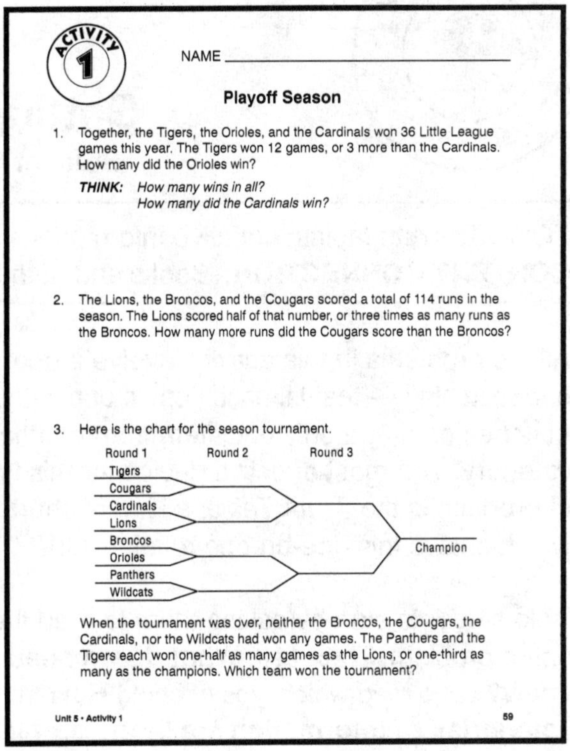

Unit 5 • Activity 1   53

ACTIVITY 2

Getting into Fiction
Student Book, pages 61 – 62

FOCUS: Truth tables; negative information
CONTENT CONNECTION: Books and authors (LANGUAGE ARTS)

All the problems in this activity involve a one-on-one matching — that is, each item in one category matches one, and only one, item from another category. The most effective device for this type of problem is the *Truth Table*, a kind of chart which shows this one-on-one relationship.

Help students start by asking them to read the entire problem. Then ask, **What do we need to know?** (who read which type of book) **How many categories of information are in the problem?** (two — readers' names, and the types of fiction)

Suggest that students construct a Truth Table:

	Sci Fi	Mys	Rom	Hist
Pearl				
Dennis				
Eve				
Alonzo				

Now, as a student reads each sentence in the problem, discuss and record the information each contains. For example,

1) A boy is reading historical fiction.

   Do we know which boy? (no) **Is either of the girls reading historical fiction?** (no)

   To record this clue, suggest that students use an X to indicate what they know is NOT true:

	Sci Fi	Mys	Rom	Hist
Pearl				X
Dennis				
Eve				X
Alonzo				

They have now eliminated two squares in the chart. Then, discuss the next sentence:

2) Pearl not reading mystery; girl (not Pearl) reading Sci Fi.

   Can we add some X's? (Yes; neither boy is reading science fiction.) **What else?** (Pearl is not reading a mystery.) Record this information. Then ask, **Is a boy or girl reading science fiction?** (girl) **Which one?** (It can't be Pearl.) **Why not?**

	Sci Fi	Mys	Rom	Hist
Pearl		X		X
Dennis	X			
Eve				X
Alonzo	X			

(because it says, "neither Pearl nor the girl . . .") **Then who is it?** (Eve) **How do you know?** (She's the only other girl.) **Is this negative or positive information?** (positive) Suggest that students record this differently from negative information for which they have used X's. For example:

	Sci Fi	Mys	Rom	Hist
Pearl		X		X
Dennis	X			
Eve	•			X
Alonzo	X			

Now ask, **Is there any other information we can record at this point?** If no one suggests any, ask, **Is Pearl reading science fiction?** (No, Eve is.) **Is Eve reading a mystery or romance?** (no, science fiction)

How can we indicate this negative information? (with X's) Continue the chart, pointing out that whenever we record positive information, we can complete that row and column with our negative symbol. We do this because the more squares we can fill in, the closer we are to a solution.

	Sci Fi	Mys	Rom	Hist
Pearl	X	X		X
Dennis	X			
Eve	•	X	X	X
Alonzo	X			

At this point, ask if there is any other information we can record. If no one responds, ask, **What is Pearl reading?** (romance) **How do you know?** (It's the only choice left.) Record this, adding X's to fill out the column:

	Sci Fi	Mys	Rom	Hist
Pearl	X	X	•	X
Dennis	X		X	
Eve	•	X	X	X
Alonzo	X		X	

Point out that one sentence (clue 2) allowed us to add two •'s and eight X's to our Truth Table.

Now have students read and discuss the last sentence of the problem:

3) Dennis is reading a mystery.

	Sci Fi	Mys	Rom	Hist
Pearl	X	X	•	X
Dennis	X	•	X	X
Eve	•	X	X	X
Alonzo	X	X	X	

Record this information, again completing the row and column with X's. Then ask, **What is Alonzo reading?** (historical fiction) **How do we know?** (It's the only choice left.)

Problems 2 and 3 can be solved in a similar fashion. (From this point on, the solutions will include a summary of the clues. Each chart will indicate the number of the clue that allows us to record either positive or negative information.)

Problems 4 and 5 are more complex. In each, three different types of information must be connected. (In Problem 4, type of fiction, setting, occupation of lead character; in Problem 5, country, plot event, and century.) For these problems, students will need this type of chart:

	W	P	N	A	US	Can	Ire	Mex
H								
R								
SF								
M								
US				
Can				
Ire				
Mex				

Note that in labeling rows and columns (as illustrated here for Problem 4), one type of information — setting — has been included in two places. That is done because in a complex Truth Table problem like this, some clues will link type of fiction to setting, while others will link setting to occupation, and still others link occupation to type of fiction. The chart allows us to record all three types of clues.

Note: It is purely arbitrary which type of headings are repeated. Either of the other two types could also have been chosen.

This type of chart requires more complex linkages of information, and therefore, more complex inferences. Guide students step-by-step through Problem 4, and let them try to work out Problem 5 on their own.

SOLUTIONS

1. Pearl, romance; Dennis, mystery; Eve, science fiction; Alonzo, historical fiction. (See explanation above for clues.)

2. Nora, *Charlotte's Web*;
   Paulette, *James and the Giant Peach*;
   Webster, *The Wind in the Willows*;
   Chun-li, *Willie Wonka and the Chocolate Factory*;
   Barton, *The Cricket in Times Square*.

	CW	JGP	WW/CF	CTS	WW
Nora	•	X①	X④	X①	X
Paulette	X③	•	X④	X③	X②
Webster	X③	X	X④	X③	•
Chun-li	X④	X④	•④	X④	X④
Barton	X	X	X④	•	X

Clues
1. Does not like JGP or CTS.
2. WW not her favorite.
3. Neither has read CW or CTS.
4. His favorite. Process of elimination determines the remaining four's. Then there is only one empty box for Paulette under JGP so that must be her favorite. X's are added to the other boxes under JGP; there is only one empty box next to Webster's name. So his favorite must be *Wind in the Willows,* and that means X's can be added to the other boxes under WW. That leaves Nora with *Charlotte's Web* as her favorite which must mean Barton's favorite is CTS. This process of elimination is indicated in the chart.

3. Arlene, Lewis Carroll; Wai, Laura Ingalls Wilder; Sloane, A. A. Milne; Wendell, R. L. Stevenson; Julio, Louisa May Alcott.

	LMA	AAM	LIW	LC	RLS
Arlene	X②	X③	X②	•	X④
Wai	X①	X③	•	X①	X④
Sloane	X③	•③	X③	X③	X③
Wendell	X④	X③	X④	X④	•④
Julio	•	X③	X	X	X④

Clues
1. Did not report on either.
2. Never read one of *his* books; therefore, did not choose any of the female authors.
3. Reported on Milne.
4. Did not report on Alcott, Wilder, or Carroll; must have reported on Stevenson. Process of elimination determines the rest.

4. Historical, writer, Ireland; romance, painter, Mexico; science fiction, nurse, United States; mystery, actor, Canada.

	W	P	N	A	US	Can	Ire	Mex
Historical	•	X	X②	X	X⑤	X⑤	•	X⑤
Romance	X⑤	•	X②	X	X⑤	X⑤	•	X⑤
Sci Fiction	X⑤	X⑤	•⑤	X③	•⑤	X③	X④	X⑤
Mystery	X	X	X⑤	•	X⑤	•⑤	X④	X⑤
US	X①	X①	•	X③				
Can	X③	X③	X③	•				
Ire	•	X	X③	X③				
Mex	X⑤	•	X③	X③				

Clues
1. Neither in U.S.
2. Not in either type.
3. Actor in Canada, but not in Sci Fi. By inference, Sci Fi cannot be Canada, since actor, who is in Canada, is not in Sci Fi. By elimination, nurse must appear in story set in U.S. (only choice left)
4. Neither set in Ireland.
5. Set in Mexico. By elimination, Sci Fi must be set in U.S. (only choice left) Since we know (Clue 3) that nurse is in U.S., nurse must be in Sci Fi story.

By elimination, we can place the mystery in Canada. Since we know the actor is in Canada (Clue 3) we can place the actor in the mystery. That leaves the writer in the historical book, and the painter in the romance, set in Mexico.

5. Voyage, England, 18th; revolution, France, 19th; war, Russia, 20th; treasure, Spain, 17th.

Clues
1. Not set in Spain or Russia; happens in 18th century. Therefore, stories in Spain and Russia are not 18th century.
2. Neither treasure nor revolution in England.

	V	R	W	T	17	18	19	20
France	X	•	X⑤	X④	X④	X	•	X③
Russia	X①	X⑤	•⑤	X④	X③	X①	X③	•③
England	•	X②	X⑤	X②	X④	•④	X	X③
Spain	X①	X④	X④	•④	•④	X①	X④	X③
17th	X①	X③	X④	•④				
18th	•①	X①	X①	X①				
19th	X	•③	X③	X③				
20th	X①	X③	•④	X④				

3. Russian story in 20th century, revolution story in 19th century.
4. Treasure set in Spain, before 19th century must, therefore, be 17th century. (only other choice)
5. War story not England or France; must be Russia. Russia story (Clue 3) in 20th century; therefore, war story is 20th century.

By elimination, the 19th century revolution must be in France; England must be voyage, which (Clue 1) is 18th century.

Unit 5 • Activity 2   57

ACTIVITY 3

Getting Around
Student Book, pages 63 – 64

FOCUS: Mutually exclusive information
CONTENT CONNECTION: Transportation (SOCIAL STUDIES)

Like the previous activity, this activity also contains problems which require us to match information. One characteristic of this information is that it is *mutually exclusive* — what is true cannot also be false; what is false cannot also be true.

A second characteristic is that the clues are *paired*; that is, each pair links an element in two categories. The problems, however, contain more than two categories.

In Problem 1, for example, the categories are *captain*, *destination* and *size of fleet*. Some clues pair captain with the size of the fleet, others pair the captain with the destination, and still others pair the destination with the size of the fleet. Students may need help understanding that if we have linked a destination and a captain from one clue, and that same captain with the size of the fleet from another clue, we have also, because of the cross-pairing, linked that destination with that fleet.

Ask students to draw a chart for Problem 1; then discuss which way they arranged the information. They may have put destinations, fleets, or captains, across the top or down the side. Suggest that, only for the purposes of discussion, you would like them to arrange their chart with the destinations and fleets across the top and the captains and fleets down the side as shown below. Put the chart on the board and do the problem aloud using the following thinking process:

	Sic	Mal	Rho	17	5	8
Philistes		X				X
Xerxes	X				X	
Mylos				X		
17			X			
5						
8		X				

Okay, now to fill in the chart, I'll enter all the information given in the problem.

1. Captain Philistes was not heading for Malta, nor did he have 8 ships in his fleet; so I'll put X's in his row under Malta and 8. I'll also put an X in the column under Malta in the 8 box because the fleet of 8 ships was not going to Malta.
2. The fleet of 17 ships did not go to Rhodes, so I'll put an X in the column under Rhodes in the 17 box.
3. Xerxes commanded a fleet of 5 ships, so I'll put an X under 5 in the row for Xerxes. But he did not go to Sicily, so I'll put an X in his row under Sicily. Captain Mylos did not command the largest fleet, so I'll put an X in his row under 17.

Ask the students if you recorded positive and negative information differently. If no one points out your error, ask the students to look for positive and negative information in the problem. Point out that it is very important to remember the following strategy step. Write the step on the chalkboard:

Step 1: Ask yourself if each piece of information yields a positive clue which is marked with a dot or a negative clue which is marked with an X.

Ask students to reread the problem to find your mistake. It is that the X under the 5 in the row for Xerxes should be a dot. Then ask, **What possibilities does this positive clue create?** (We can put X's in the 17 and 8 boxes in the row for Xerxes and under 5 in the row for Philistes and Mylos.) Ask students to tell you what these X's now show you because of the boxes that remain empty. (The only empty number box in Philistes' row is under 17; so we can put a dot there. The only empty number box in Mylos' row is under 8; so we can put a dot there.) Write the following on the chalkboard:

Step 2: Fill in the X's to show what cannot be true.

Ask, **If we know that Philistes had a fleet of 17 ships and our chart shows that 17 ships did not go to Malta, what can we deduce?** (that Philistes did not go to Malta so he must have gone to Sicily)

Ask the students what to do after placing the dot under Sicily. (We can place X's in the Sicily column for Xerxes and Mylos and a dot in the Sicily column for 17 with X's for 5 and 8. We also can place an X under Malta in the 17 row.)

Now, to find whether Xerxes or Mylos went to Malta, what do we do? If students are stuck, write the following on the chalkboard:

Step 3: Reread the problem to see what information can be reused.

The problem states that one fleet headed for Malta and another set sail with 8 ships, so we know that 8 ships did not go to Malta. Since Xerxes had 8 ships, he did not go to Malta; so he must have gone to Rhodes. This leaves Mylos as the one who went to Malta.

Leave the strategy steps on the chalkboard as students try Problems 3 and 4 on their own.

Problem 5 is more easily solved using a *matching list* method of organization, rather than a truth table because it requires matching items from four categories.

Ask, **What do we need to know about each person?** (occupation, place they started from, place they are going to, means of travel)

List these headings on the board:

<u>OCCUPATION</u>   <u>LEAVING</u>   <u>GOING TO</u>   <u>BY</u>

Have a student read the first sentence of the problem. Ask, **How many different people are mentioned here?** (four, since each clue refers to a different person) **Is the business woman leaving Miami?** (no) **Is she going to Bermuda?** (no) **Is she driving a car?** (no)

Should we list all this information on the same line in our chart? (no, because each item refers to someone different) **Right! They are mutually exclusive!** Suggest that students fill in the chart in the following manner:

<u>OCCUPATION</u>   <u>LEAVING</u>   <u>GOING TO</u>   <u>BY</u>
1. businesswoman
2.   Miami
3.   Bermuda
4.   car

Ask them now to place the information from the second clue in the chart. Guide them with questions: **Can the salesman go in the first row of the chart?** (no, he's not the business woman) **Can he go in the second row?** (no, he's not leaving from Miami) **How about the third row?** (maybe; we can't tell for sure) **The fourth row?** (no, he's not driving) **Therefore, which row must he go in?** (third)

Use similar questions to help students with the rest of the clues.

Note: This same type of chart, or matching list, can also be used for the first four problems of the activity. Some students may find it an easier device than truth tables.

Unit 5 • Activity 3

SOLUTIONS

1. Mylos went to Rhodes with 8 ships.

2. Xerxes went to Malta; Philistes went to Sicily; and Mylos went to Rhodes.

	Sic	Mal	Rho	17	5	8
Philistes	•	X①	X③	•	X③	X①
Xerxes	X③	•	X③	X③	•	X③
Mylos	X③	X③	•	X③	X③	•
17	•	X③	X②			
5	X③	•	X③			
8	X③	X①	•			

Clues

1. Philistes did not go to Malta; did not have 8 ships; 8 ships did not go to Malta.
2. 17 ships did not go to Rhodes.
3. Xerxes had 5 ships; by elimination, Philistes had 17 and Mylos, 8; Xerxes did not go to Sicily; therefore, 5 ships did not go to Sicily. Since Philistes' 17 ships did not go to Rhodes (Clue 2), Philistes did not go to Rhodes, and must have gone to Sicily, as did his 17 ships; so, by elimination, 8 ships went to Rhodes, and 5 to Malta. Since Xerxes had 5 ships (Clue 3), he must have gone to Malta, and Mylos to Rhodes.

Note: Remind students that the process of elimination is a powerful strategy in reasoning.

3. Antonius, Egypt, 5 legions;
   Messina, Britain, 3 legions;
   Lucius, Spain, 4 legions;
   Nestor, Gaul, 6 legions.

Clues

1. Antonius did not go to Britain; 4 legions did not go to Britain; Antonius did not lead four legions.
2. Messina did not have 6 legions and did not go to Gaul; 6 legions went to Gaul.
3. Lucius did not go to Egypt or Britain.
4. 4 legions went to Spain; Nestor did not go to Spain or lead 4 legions; since

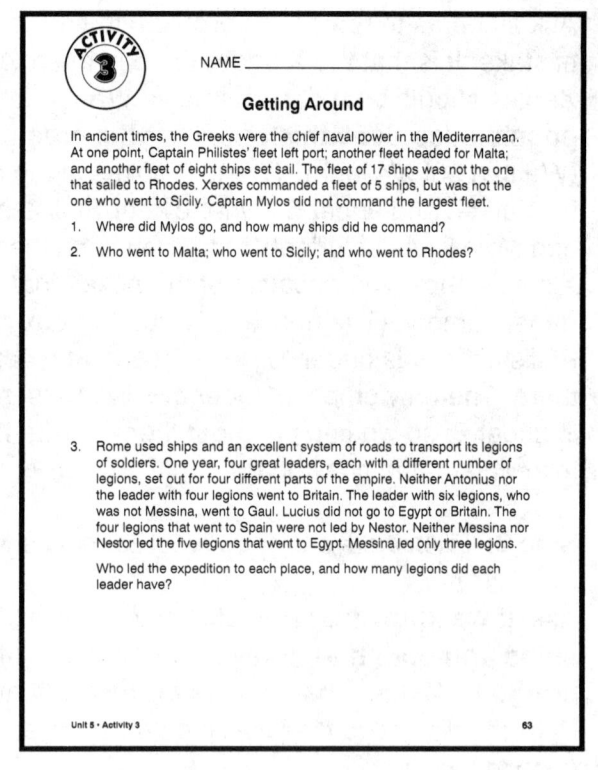

ACTIVITY 3

NAME _____

Getting Around

In ancient times, the Greeks were the chief naval power in the Mediterranean. At one point, Captain Philistes' fleet left port; another fleet headed for Malta; and another fleet of eight ships set sail. The fleet of 17 ships was not the one that sailed to Rhodes. Xerxes commanded a fleet of 5 ships, but was not the one who went to Sicily. Captain Mylos did not command the largest fleet.

1. Where did Mylos go, and how many ships did he command?
2. Who went to Malta; who went to Sicily; and who went to Rhodes?

3. Rome used ships and an excellent system of roads to transport its legions of soldiers. One year, four great leaders, each with a different number of legions, set out for four different parts of the empire. Neither Antonius nor the leader with four legions went to Britain. The leader with six legions, who was not Messina, went to Gaul. Lucius did not go to Egypt or Britain. The four legions that went to Spain were not led by Nestor. Neither Messina nor Nestor led the five legions that went to Egypt. Messina led only three legions.

Who led the expedition to each place, and how many legions did each leader have?

	B	G	E	S	4	6	5	3
Antonius	X①	X⑤	•	X①	X①	X⑤	•	X⑤
Messina	•	X②	X⑤	X⑥	X⑥	X②	X⑤	•
Lucius	X③	X⑥	X③	•	•	X⑥	X⑤	X⑤
Nestor	X⑤	•	X⑤	X④	X④	•	X⑤	X⑤
4	X①	X②	X④	•				
6	X②	•	X②	X②				
5	X⑤	X②	•	X④				
3	•	X②	X⑤	X④				

Antonius (Clue 1) did not lead 4 legions, he didn't go to Spain, where the 4 legions went.

5. Messina did not go to Egypt or lead 5 legions; Nestor did not go to Egypt or lead 5 legions; 5 legions went to Egypt. By elimination, Antonius must have gone to Egypt, taking 5 legions. That leaves 3 legions for Britain, with Nestor going to Gaul with 6 legions. (from Clue 2)

6. Messina led 3 legions, so by elimination, Lucius led 4 legions. Since 4 legions (Clue 4) went to Spain, Lucius went to Spain, leaving Messina going to Britain.

4. Venice, China, silk;
   Germany, Baltic, fish;
   Portugal, East Indies, spice;
   Morocco, Timbuktu, gold.

	V	G	P	M	gold	fish	spice	silk
Timbuktu	X③	X②	X①	•④	•①	X①	X①	X①
Baltic	X②	•②	X②	X②	X①	•②	X②	X②
China	•③	X③	X③	X③	X③	X③	X③	•③
E. Indies	X③	X②	•④	X④	X①	X②	X③	X③
gold	X③	X②	X①	•③				
fish	X②	•②	X②	X②				
spice	X③	X②	•③	X③				
silk	•③	X②	X③	X③				

Clues
1. Gold came from Timbuktu; Portugal did not trade in gold, not with Timbuktu.
2. Germany, fish, Baltic go together.
3. Spice not to Venice; Venice traded with China; spice not Indies and silk from China; also, Portugal must trade with East Indies, Morocco with Timbuktu.
4. Morocco and East Indies did not trade, neither produced silk.

Since China traded in silk and also traded with Venice, Venice traded in silk. The gold went to Morocco, and the spice to Portugal.

5.

	OCCUPATION	LEAVING	GOING TO	BY
1.	businesswoman	New York	Washington	train
2.	florist	Miami	Bahamas	ship
3.	salesman	Boston	Bermuda	plane
4.	college student	Seattle	St. Louis	car

Clues
1. Businesswoman, Miami, Bermuda, and car are all different; therefore, all go in different rows.
2. Salesman can't go in row 1 (not a businesswoman) or row 2 (not leaving Miami) or row 4 (not driving) so he must go in row 3.
3. College student leaving from Seattle cannot go in row 2 (Seattle, not Miami) so must go in row 4.
4. Bahamas can't be row 3 (not Bermuda) or row 4 (ship, not car); postpone — could be either row 1 or row 2.
5. New York – Washington must go in row 1; only row left with space for origin and destination. So Bahamas and ship must go in row 2 (only row with both spaces open.)
6. By elimination, the florist is row 2, and St. Louis, row 4.
7. Boston must be row 3; train, row 1. Both logic and common sense tell us the plane is going to Bermuda.

Unit 5 • Activity 3

4. In the Middle Ages, Venice, Germany, Portugal, and Morocco were all important centers of trade. Portugal did not trade in the gold that came from Timbuktu. Germany traded in fish with the Baltic region. The spice trade did not go to Venice, which traded with China. Morocco did not trade with the East Indies, which did not produce the silk that was traded.

Which trading center traded with each area? What goods did each trading center handle?

5. Recently, a business woman, a person leaving Miami, someone going to Bermuda, and a person driving a car all left home on the same day. The salesman is not going to drive, and he is not the one leaving from Miami. The college student is leaving from Seattle, but is not the one taking a train or plane. Someone is taking a ship to the Bahamas. The person making the New York – Washington trip is not the one going by plane. The florist is not going to St. Louis. The person taking the train is not leaving from Boston.

Match each occupation with the person's home town, destination, and means of travel.

ACTIVITY 4

Environments
Student Book, pages 65 – 66

FOCUS: Practice in logical reasoning
CONTENT CONNECTION: Climate regions (SCIENCE)

The students by now should be fairly adept at organizing information, recognizing clues, and drawing logical conclusions. The problems of this activity are designed to give them extended practice in solving logic problems and verbalizing their thinking processes.

Ask students to work in cooperative teams to solve these problems. To make sure that each student understands the process, you might want each team member to write out the thinking steps he or she used, or to be prepared to explain them aloud to the class.

Students might also like to experiment with the organizing technique of the parallel list introduced in the last activity. Some may find this a more effective strategy. It is especially appropriate for Problem 4, which requires matching items in four categories.

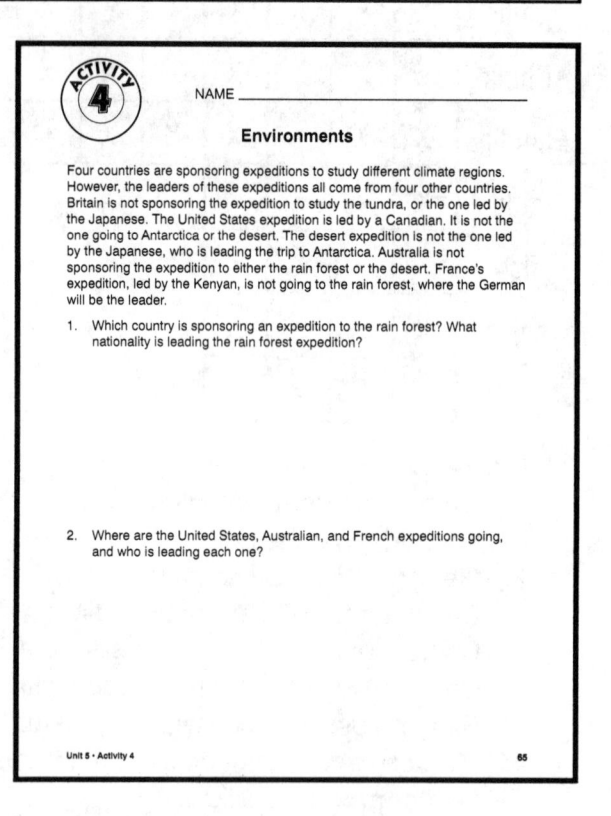

SOLUTIONS

1. Britain, rain forest, German

2. U.S., tundra, Canadian;
   Australia, Antarctica, Japanese;
   France, desert, Kenyan.

ACTIVITY 4
Environments

Four countries are sponsoring expeditions to study different climate regions. However, the leaders of these expeditions all come from four other countries. Britain is not sponsoring the expedition to study the tundra, or the one led by the Japanese. The United States expedition is led by a Canadian. It is not the one going to Antarctica or the desert. The desert expedition is not the one led by the Japanese, who is leading the trip to Antarctica. Australia is not sponsoring the expedition to either the rain forest or the desert. France's expedition, led by the Kenyan, is not going to the rain forest, where the German will be the leader.

1. Which country is sponsoring an expedition to the rain forest? What nationality is leading the rain forest expedition?

2. Where are the United States, Australian, and French expeditions going, and who is leading each one?

Clues
1. Not sponsoring tundra; not led by Japanese.
2. Led by Canadian.
3. Not to Antarctica or desert; therefore, Canadian not going to Antarctica or desert.
4. Japanese not going to desert; leading trip to Antarctica.
5. Not sponsoring rain forest or desert.
6. Led by Kenyan; not to rain forest; German going to rain forest; Japanese must lead Australia's expedition (only one left); Japanese going to Antarctica (Clue 4); therefore, Australia sending expedition to Australia. Canadian must be going to tundra (only spot left) and Kenyan to desert; therefore, France's expedition to desert, Britain's, rain forest, led by German; and U.S., tundra.

3. Desert, Australia;
   rain forest, South America;
   mountains, Asia;
   plains, Africa;
   woodlands, North America.

	d	rf	m	p	w
Asia	X²	X²	•	X²	X²
Africa	X³	X³	X²	•	X¹
N. Amer.	X²	X²	X²	X²	•
S. Amer.	X³	•	X²	X²	X¹
Australia	•	X³	X²	X³	X¹

Clues
1. Not study woodlands in Africa, South America, or Australia.
2. Mountains in Asia; by elimination, woodlands in North America.
3. Desert not in South America or Africa; must be Australia.
4. Plains not in South America, therefore, must be in Africa; by elimination, rain forest must be South America.

4.

Prize	First Name	Last Name	Topic
3	Martin	Ramirez	Arctic
5	Sam	Ching	deserts
1	Mary	Takeshima	coral reefs
2	Paul	King	rain forests
4	Valerie	Williams	seashore

Clues
1. 3rd prize, deserts, Mary, Paul, Williams all go in different rows; they are different people.
2. Postpone, cannot use yet.
3. Martin Ramirez must go in row 1. (one of two rows with space for first and last name; other is not possible because Ramirez didn't write about deserts.) King, 2nd prize, and rain forests must now go with Paul (one of two rows with those three spaces available; cannot go with Mary because King is a man. (Clue 2)
4. Valerie must go with Williams; is not Mary and did not write about deserts.
5. Takeshima, 1st prize, coral reefs must go with Mary. (only space available for all three)
6. Deserts got 5th prize; Arctic book must go in first row; desert book author must be Sam Ching; Valerie therefore got fourth prize.

5. Dwyer, geology, Serengeti;
   Assad, animals, Amazon;
   Clay, plants, Pacific;
   Burns, climate, Atacama.

	S	At	Am	Pac	an	geol	pl	cl
Dwyer	•④	X③	X②	X④	X①	•④	X④	X③
Assad	X②	X①	•②	X①	•①	X①	X①	X①
Clay	X④	X③	X②	•④	X①	X④	•④	X③
Burns	X②	•③	X②	X③	X①	X②	X③	•①
animals	X②	X②	•②	X②				
geology	•②	X②	X②	X②				
plants	X②	X③	X②	•②				
climate	X②	•③	X②	X③				

Clues

1. Assad, animals; not Pacific or Atacama.
2. Not Assad or Burns; Assad must be studying Amazon (only one left); therefore, Amazon goes with animals, his specialty (Clue 1); Serengeti must go with geology.
3. Burns must be studying Atacama, studies climate; by elimination, plants must be studied in Pacific.
4. Clay is not a geologist; must be plant specialist; therefore, Dwyer is geologist; goes with Serengeti (Clue 2); Clay must be Pacific.

---

3. Teams of scientists are going to Asia, Africa, North America, South America, and Australia to study different geographical regions: desert, rain forest, mountains, plains, and woodlands. They will not study woodlands in Africa, South America, or Australia. In Asia, they will study mountains. The desert under study is not in South America or Africa. They will not go to South America to study plains.

   On which continent will they study each type of region?

4. Every year, the Environmental Society awards five prizes for the best books about the environment. At this year's Awards Dinner, the third-prize winner sat with the author of a book on deserts, and Mary, Paul, and Ms. Williams. Mr. King, whose name is not Sam, won second prize for his book about rain forests. Martin Ramirez did not write about deserts. Valerie wrote about the seashore. Ms. Takeshima won first prize for her book about coral reefs. Fifth prize went to the book about the deserts written by Mr. Ching, not the one about the Arctic.

   What is the first and last name of each author; what prize did each win; and what was the topic of each book?

5. Four scientists, Dr. Dwyer, Dr. Assad, Dr. Clay, and Dr. Burns, are specialists in animal life, geology, plant life, and climate, but not necessarily in that order. Among them, they are presently studying the Serengeti Plains, the Atacama Desert, the Amazon River Delta, and a Pacific Atoll. Dr. Assad, whose specialty is animals, is not studying a Pacific Atoll or the Atacama Desert. The scientist studying the Serengeti Plain, a specialist in geology, is not Assad or Burns. Dr. Burns, who is not going to a Pacific Atoll, studies climate. Dr. Clay is not a specialist in geology.

   What is each scientist's specialty, and what region is he or she studying?

Activity 5

Challenge
Student Book, pages 67 – 68

FOCUS: Review

These problems can serve as an assessment of students' understanding; or they could be assigned to small groups of students as additional practice. You may want to assign tasks to each group member to promote cooperative learning.

Solutions

1. Rosa, soccer; Eduardo, hockey; Ana, tennis; Carlos, swimming; Juan, basketball.

	Sw	T	H	So	B
Rosa	X⁵	X⁵	X³	•	X⁴
Eduardo	X¹	X¹	•	X⁵	X⁴
Ana	X⁵	•	X³	X²	X⁴
Carlos	•	X³	X⁵	X²	X⁴
Juan	X⁴	X³	X⁴	X⁴	•

Clues
1. Does not swim or play tennis.
2. Neither plays soccer.
3. Boys do not play tennis; girls do not play hockey.
4. Plays basketball.
5. Soccer player is girl, but not Ana (Clue 2) so it must be Rosa. Therefore, Ana must play tennis (only sport left), Carlos is the swimmer, and Eduardo, the hockey player.

2.

School	Street	Team	Color
North	Main	Tigers	blue
East	Broadway	Bears	green
South	Central	Lions	red
West	Grand	Mustangs	orange

Clues
1. North, Broadway, Lions, orange all go in different rows.
2. Postpone; cannot use yet.
3. Tigers cannot go in row 4 (do not use orange) or 2 (not on Broadway) or 3 (they are not the Lions); they must go in row 1 and go to North School; Tigers do not use green, postpone.
4. Grand goes in row 4, with orange; must be West or South. East School must be in row 2 with Broadway, since neither West or South (Clue 2) is on Broadway.
5. Central does not go in row 1 with North, so it must go in row 3. That means the school in row 3 is not West. (West is not on Central.) So it must be South, and West must go in row 4.
6. The Bears and their color, green, must go in row 2 (the only row with a space for both team and color); Main Street can only go in row 1.
7. The Mustangs must go in row 4; blue (since it does not go with South) must go in row 1, leaving red for row 4.

3. 11,000,000

   $13 + 16 = 29$
   $ +13$
   $\overline{42}$

   Cal 53
   Fla −42
   $\overline{11}$ Ohio

4. Wandalee, Snoopy, canary; Kumi, Chum, cat; Bill, Ariel, dog; Marco, Rover, snake.

	R	A	S	C	dog	cat	can.	snake
Wandalee	X¹	X³	•³	X³	X¹	X²	•³	X³
Kumi	X⁴	X²	X²	•⁴	X²	•²	X²	X²
Bill	X⁴	•⁴	X³	X⁴	•⁴	X²	X³	X⁴
Marco	•⁴	X⁴	X³	X⁴	X⁴	X²	X³	X⁴
dog	X¹	•⁴	X⁴	X⁴				
cat	X⁴	X²	X²	•⁴				
canary	X⁴	X⁴	•⁴	X⁴				
snake	•⁴	X⁴	X⁴	X⁴				

   Clues
   1. does not own dog, or Rover. Rover is therefore not the dog.
   2. Owns cat; does not own Ariel or Snoopy; Ariel and Snoopy are not cats.
   3. Neither owns canary; therefore Wandalee owns canary; neither Bill nor Marco owns Snoopy or Chum, so Wandalee (only one left) owns Snoopy.
   4. Rover is the snake; therefore, the cat is Chum, owned by Kumi (Clue 2); Rover is not Bill's, so he must be Marco's snake, leaving Bill with a dog named Ariel.

5. Take one out of the bag mismarked tulips and daffodils.

	T/D	T	D	
	T/D	T	D	(wrong labels)
If tulip, then	T	D	T/D	(correct labels)
If daffodil, then	D	T/D	T	(correct labels)

Strategy Discussion

This unit contains several vocabulary items which may be new to students. Ask them to brainstorm definitions for *negative information*, *positive information*, *mutually exclusive information*, *inference*, *clue*, and *truth table*.

Different teams or small groups could select a term to define, and then share their definition with the class. They might also cite examples in this unit of each of these terms.

Extend this activity by asking students to work in groups to create a definition of logical reasoning. Have each group present its definition to the class. A committee of *editors* — one person from each group — could work cooperatively to combine these definitions into one. Volunteers could make a poster of this *class definition*.

Extension

Problems of this type are difficult to construct, but some students might like to challenge the class with original problems. Encourage interested students to *edit* each other's problems to be sure that they have provided enough, but not too many, clues.

You might suggest that students find, and try, puzzles of this type in popular "logic puzzle" magazines.

All students could look for, and bring to class, examples of logical reasoning, which they find in such sources as textbooks, newspapers, and magazines.

Use Logical Reasoning
Student Book, pages 69 – 70

This page of the student text is designed to be taken home as a family involvement activity.

Included for the parent is a brief explanation of the unit content with a summary of the strategy which has been developed.

Along with that background information, there are several problems which the student and the family can solve together.

This page may be assigned as soon as Activity 4 of the unit has been completed.

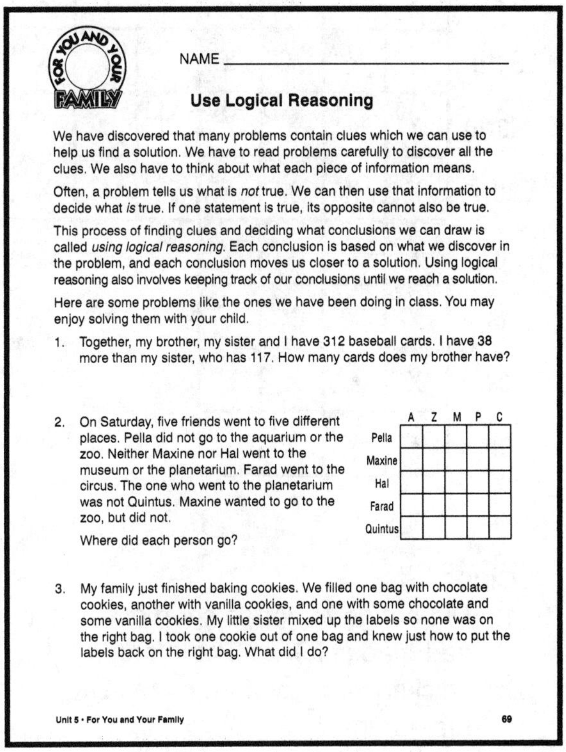

Solutions

1. The brother has 40 cards.

   Sister   117   312
   I   155 (117 + 38)   − 272
                                  40

2. Pella, planetarium; Maxine, aquarium; Hal, zoo; Farad, circus; Quintus, museum.

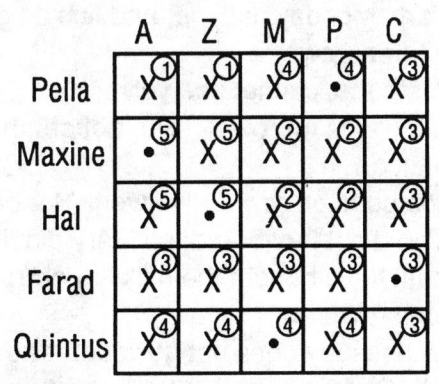

	A	Z	M	P	C
Pella	X¹	X¹	X⁴	•⁴	X³
Maxine	•⁵	X⁵	X²	X²	X³
Hal	X⁵	•⁵	X²	X²	X³
Farad	X³	X³	X³	X³	•³
Quintus	X⁴	X⁴	•⁴	X⁴	X³

Clues
1. Did not go to aquarium or zoo.
2. Neither went to museum or planetarium.
3. Went to circus.
4. Quintus did not go to planetarium. Therefore, must have been Pella. (only one left) That leaves only Quintus for the museum.
5. Maxine did not go to zoo; must have gone to aquarium, leaving only the zoo for Hal.

3. Take a cookie from the bag wrongly marked "Chocolate and Vanilla."

	C/V	C	V	
	C/V	C	V	(wrong labels)
If C removed, then	C	V	C/V	(correct labels)
If V removed, then	V	C/V	C	(correct labels)

Unit 5 • For You and Your Family

4. Tao, cafeteria, fried chicken; Seth, coffee shop, bacon/lettuce/tomato; Mario, restaurant, chef's salad; Laurie, snack bar, tuna salad.

	CS	R	C	SB	fc	ts	cs	blt
Tao	X¹	X¹	•⁴	X⁴	•¹	X¹	X¹	X¹
Seth	•⁵	X⁵	X²	X⁵	X¹	X²	X⁵	•⁵
Mario	X⁵	•⁵	X²	X⁵	X¹	X²	•⁵	X⁵
Laurie	X⁵	X⁵	X⁴	•⁵	X¹	•⁴	X⁴	X⁴
fc	X¹	X¹	•⁴	X⁴
ts	X⁵	X³	X⁴	•⁵
cs	X³	•³	X³	X³
blt	•⁵	X³	X⁴	X⁵

Clues

1. Did not go to coffee shop or restaurant; had fried chicken. Therefore, fried chicken was not from coffee shop or restaurant.
2. Neither had tuna; neither went to cafeteria.
3. Chef's salad from restaurant.
4. Had tuna; did not go to cafeteria. Tao (only one left) went to cafeteria. Therefore chicken, which Tao had, came from cafeteria.
5. Did not go to snack bar; had BLT. Mario (only one left) had chef salad. Since chef salad was from restaurant (Clue 3), Mario went to restaurant. Therefore Seth (only one left) ate at the coffee shop, and Laurie went to the snack bar. We know (Clue 4) that she had tuna, therefore tuna goes with snack bar. That leaves the BLT for the coffee shop where Seth ate.

5.

Person	Day	Item
Mary	Saturday	bread
Bob	Monday	milk
Lucy	Wednesday	lettuce
George	Thursday	fish

Clues

1. Mary, Monday, lettuce, and fish all go on different rows.
2. Postpone; cannot use yet.
3. Lucy goes on row 3. (She bought the lettuce.)
4. Bob goes on row 2. (He went Monday.) He did not buy bread; so Mary did. He must have bought the milk. George must go in row 4.
5. Wednesday does not go with bread, or with George. (Clue 2) So it must go with Lucy. Since Saturday doesn't go with George (Clue 2) it must go with Mary, leaving Thursday for George.

UNIT 6

STRATEGY: Set Up an Equation

INTRODUCTION

Many of the strategies introduced in *Problem Solving Connections* are organizing strategies. That is, they are methods of arranging the information given in a problem so that we can readily access this information as we seek a solution. *Make a List, Table, or Chart* is one such strategy.

Writing an equation is another useful organizing strategy. This unit introduces students to simple equations as a means of representing the information in a problem.

Briefly stated, we are teaching students that algebraic equations use variables for the unknown number and symbols for the operations. Thus $12N - 3 = 93$ is an equation meaning "twelve times some number minus three is 93." The unknown is the number which must be multiplied by 12. The letter N is a variable; in another problem, such as $2N = 6$, the same letter N has an entirely different value. It is important that students are aware that the letter chosen to represent, or stand for, the unknown may be any letter. N is often used because it reminds us of "number."

Many algebra problems use X to stand for the unknown. The equation $2X = 6$ means the same thing as $2N = 6$, or $2Y = 6$, or $2B = 6$.

The goal of this unit is to develop students' abilities to recognize that they can represent data with equations and use those equations to solve problems.

OBJECTIVES

Students will:
- construct equations from data in problems.
- determine the value of the variable in an equation.
- use equations to solve problems.
- create problem statements based on equations.

OVERVIEW

Activity 1 Recognizing equations.
Activity 2 Setting up equations.
Activity 3 Creating problems from equations.
Activity 4 Solving area, perimeter, and volume problems with equations.
Activity 5 Review of problem types.

When students have completed this unit, you may wish to assign the Expert Problems, Set E, on pages 93-94, for cumulative strategy application.

ACTIVITY 1

Using Equations
Student Book, pages 73 – 74

FOCUS: Identifying equations; basic vocabulary
CONTENT CONNECTION: Algebraic notation (MATHEMATICS)

This activity reviews the basic concept of an equation, and provides examples of simple equations. The amount of time which students will need to spend on this activity depends on the extent of their familiarity with equations.

This can be a brief review for students already familiar with using equations. Students less experienced with equations will need to spend more time reflecting on and discussing each problem.

Problem 1 uses a balance scale to introduce the concept of an equation. Most students will have little difficulty recognizing that to make it balance, one must add 5 cubes to the higher side.

MANIPULATIVES: *A real balance scale, if available, along with 14 identical objects to weigh, would give students a chance to balance an equation in a "hands-on" manner.*

In Problem 2, help students recognize that an *equation* balances. That is, what is on the left of the equal sign balances, or equals, or is the same as, what is on the right of the equal sign.

Explain that $2 + 5 = 7$, an equation, is a way of describing what they did to balance the scale in Problem 1.

The simple answer to Problem 3 is that N stands for 5, the number we need to add to 2 to equal 7. It is also the number of cubes we have to put on the higher side of the scale to make it balance.

Use this example to explain that in an equation, a letter, like N, may be used to stand for something *unknown*, something we have to figure out. Use the familiar question **What do we need to know?** With the equation $2 + N = 7$, we need to know the *value* of N — or what number N stands for.

Give another example, such as $3 + N = 12$ and ask, **What do we need to know?** (the value of N, or what N stands for) Ask students to solve the equation. Then ask, **Did N in this equation have the same value as N in the previous equation?** (No; this time it was 9, last time it was 5.)

Point out that the value of an unknown such as N is *variable* — that is, its value changes from equation to equation.

Problem 4 provides practice in recognizing the basic structure of an equation, and in verbalizing the process for solving one.

It is important for students to verbalize the solution process if they are to internalize the relation between equations and the operations they symbolize.

Identifying whether N (or any variable in an equation) is greater than or less than the number to the right of the equal sign helps students to focus on that relationship. In both Problem 4 and Problem 5, students should explain why N must be greater than, or less than, the number to the right of the equal sign.

Problem 5 extends this practice with a variety of operations, including fractions.

MANIPULATIVES: *To emphasize the parallel between equations and the balance scale, you might want to have students make a "balance scale" model of each equation they solve.*

SOLUTIONS

1. Put five more cubes on the higher side because there are seven on one side and you need seven on the other side. To balance, both sides must be the same, as with the equal sign.

2. a. equal   d. false
   b. false   e. equal
   c. equal   f. equal

3. N stands for 5, because 2 + 5 = 7.

4. a. N = 17
      N is greater than 12.
      Find the number that is 5 greater than 12.
   b. N = 7
      N is less than 14.
      Find what number times 2 equals 14.

5. a. N is greater than 4.
      N = 21
      Find what number is 4 greater than 17.
   b. N is less than 20.
      N = 5
      Find what number times 4 equals 20.
   c. N is less than 32.
      N = 7
      Find what number plus 25 equals 32.
   d. N is greater than 43.
      N = 50
      Find what number is 7 greater than 43.
   e. N is less than 1.
      N = 1/2
      Find what number plus 1/2 equals 1.
   f. N is less than 1.
      N = 1/4
      Find what number times 4 is equal to 1.

EXTENSION

For more able students, use the items in Problem 5 to demonstrate solving an equation by performing the same operation on both sides of the equal sign. For example:

a. $N - 17 = 4$
   $N - 17 + 17 = 4 + 17$
   $N = 4 + 17$
   $N = 21$

b. $4N = 20$
   $\dfrac{4N}{4} = \dfrac{20}{4}$
   $N = 5$

c. $25 + N = 32$
   $25 - 25 + N = 32 - 25$
   $N = 32 - 25$
   $N = 7$

d. $N - 7 = 43$
   $N - 7 + 7 = 43 + 7$
   $N = 43 + 7$
   $N = 50$

e. $N + \tfrac{1}{2} = 1$
   $N + \tfrac{1}{2} - \tfrac{1}{2} = 1 - \tfrac{1}{2}$
   $N = \tfrac{2}{2} - \tfrac{1}{2}$
   $N = \tfrac{1}{2}$

f. $4N = 1$
   $\dfrac{4N}{4} = \dfrac{1}{4}$
   $N = \tfrac{1}{4}$

ACTIVITY 1 — Using Equations

1. Suppose you had a balance scale . . .
   . . . and some cubes that each weigh 30 grams . . .
   and you put 7 of the cubes on one side of the balance scale and 2 cubes on the other.
   What would you have to do to make the scale balance again? Why? How is a balance scale like an equal sign (=)?

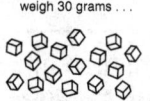

   THINK: When does a balance scale balance? What does an equal sign tell you about the numbers to the left and to the right of it?

2. The number sentence 2 + 5 = 7 is called an *equation* because both sides balance. They are equal. Which of the following equations are equal? Which ones are false statements?

   a. 3 + 18 = 21   d. 1/2 + 1/2 = 1/4
   b. 17 − 9 = 11   e. 3 × 1/3 = 1
   c. 36 + 4 = 9   f. 7 × 15 = 105

3. Another way to write the problem about the balance scale is 2 + N = 7. In this equation, what does N stand for? How do you know?

4. Here are some other equations. Solve them, and answer the questions about them.

   a. N − 5 = 12   What is N equal to?
                  Is N greater or less than 12?
                  What do you have to do to solve this equation?

   b. 2N = 14   2N means 2 times N.
                  What does N equal?
                  Is N greater or less than 14?
                  What do you have to do to solve the equation?

5. For each of the following, tell whether N is greater than or less than the number to the right of the equal sign. Then solve the problem and explain what you did to solve it.

   a. N − 17 = 4
   b. 4N = 20
   c. 25 + N = 32
   d. N − 7 = 43
   e. N + 1/2 = 1
   f. 4N = 1

ACTIVITY 2

Find the Equation
Student Book, pages 75 – 76

FOCUS: Converting information into equations
CONTENT CONNECTION: Equations (MATHEMATICS)

Before we can use an equation to solve a problem, we must first set up, or create, the equation. The equation must represent accurately the information and relationships expressed in the problem, or it will be of no value as a strategy for solving the problem.

This activity provides experience in converting, or translating, the factual information contained in problems into algebraic equations which represent those problems.

Since the concept of equations is relatively new for many students, you may want to present these problems as a whole-class activity. Start by having them read Problem 1. Ask, **What word in the problem could we replace with the word *equals*?** (is) Have a student read the sentence aloud, substituting *equals* for *is* so that students understand that both terms have the same meaning.

Then ask, **What is another way of writing the word *equals*?** (Use the "=" sign.) Draw "=" on the board. **What number should we put to the right of the "=" sign?** (52) Add this to the equation on the board:

$$= 52$$

Ask, **What is equal to 52 in the problem?** (Three times some number plus 7.) **Do we know yet what that number is?** (no) **What can we use to stand for a number we have to find?** (a letter like N) Add N to the equation, leaving a space:

$$N \quad = 52$$

Does the problem say that N equals 52? (No, 3N plus 7 equals 52.) **How do we show this in our equation?** (Add a 3 before the N and a plus 7 after the N.) Complete the equation:

$$3N + 7 = 52$$

Is 3N greater than or less than 52? (less) **How can we tell?** (because we have to add 7 to it to make 52) **How much less than 52 is 3N?** (7 less) **What can we do to find a number that is 7 less than 52?** (Subtract 7 from 52; 45.) **Is N equal to 45?** (no, 3N = 45) **What can we do to find a number that multiplied by 3 equals 45?** (We can divide 45 by 3.) **Is N equal to 15?** (yes) **How do you know?** (3 times 15 plus 7 is equal to 52.) Show this below the equation:

$$3N + 7 = 52$$
$$3(15) + 7 = 52$$
$$45 + 7 = 52$$

Problems 2 and 3 can be solved in a similar fashion. Teams of students can work on these problems and then explain their solutions to the class.

Remind students to ask each time, "Is N greater than or less than the number to the right of the equal sign?" This will help them to select the appropriate operations.

In Problem 2 students should understand that "eight less than twice some number" means 2N – 8, not 8 – 2N.

Problems 4 and 5 require careful analysis before students will be able to set up their equations. After they read Problem 4, ask, **What do we need to know?** (the number of light bulbs in a carton) **How many cartons are there?** (9) **If we knew how many light bulbs in a carton, and we had 9 cartons, how could we find how many light bulbs in all?** (multiply) **Do we know how many light bulbs in a carton?** (No, that's what we have to find out.)

Tell them that a useful first step is to identify what part of the problem should be represented by the equal sign. In Problem 4, it is the words "needed a total of."

Next, students should look for the information that goes to the right of the equal sign. In Problem 4, it is 61 (light bulbs). The partial equation is now:

$$= 61$$

The third step is to decide what numbers, and what operations, belong to the left of the equal sign. (Volunteers might summarize and make a poster of these steps for the classroom.)

Ask, **What should we use to stand for the number of light bulbs?** (a letter) Suggest that they use L to remind them that the problem is about light bulbs. **How many cartons are there?** (9) **How can we write this information?** (9L)

Does 9L equal 61? (no) **How do we know?** (still needed 7 bulbs to make 61) **Is L greater than or less than 61?** (less; because 9L plus 7 equals 61) Ask students to suggest an equation and write it on the chalkboard:

$$9L + 7 = 61$$

Then ask students to solve the equation by finding the value of L. (It is 6.)

They should now be ready to try Problem 5, a similar problem, on their own.

SOLUTIONS

1. $3N + 7 = 52$
   $N = 15$

2. $2N - 8 = 26$
   $N = 17$

3. $1/3 N + 12 = 41$
   $N = 87$

4. $9L + 7 = 61$
   $L = 6$

5. $6P + 4 = 52$
   $P = 8$

ACTIVITY 2

NAME _____

Find the Equation

1. Three times some number plus 7 is 52. What is that number?

   THINK: What is the equation for this problem?
   What operation will solve it?

For each of the following problems, write an equation and then solve it.

2. Eight less than twice some number is 26.

3. One-third of some number plus 12 is 41.

4. Roger needed a total of 61 light bulbs. He had nine cartons of light bulbs, but still needed 7 more. How many bulbs are in each carton?

5. Sophia stacked some pennies in six equal piles and had four pennies left over. Altogether, she had 52 pennies. How many pennies were in each pile?

ACTIVITY 3

Storytelling
Student Book, pages 77 – 78

FOCUS: Translating equations into story problems

To help students develop an intuitive sense of how equations are related to problems, this activity requires them to create a story problem based on a given equation.

Be sure that students understand the relationship between the equation and the problem in the example. **Why does the story problem say "three times Nora's age . . . ? "** (because the equation says "3N . . .")

Ask students to decide if this is the only story problem that could be made from this equation. (no) Allow students time to make up original problems based on this equation. As they present their problems, check to be sure that each is based on the equation 3N + 3 = 39. (N = 12)

Extension: After students have reviewed this example, ask them to write an equation that represents Nora's age and her Dad's age 10 years from now. (2N + 5 = 49 is one possibility.)

When students have completed their story problems, encourage them to challenge other students in the class to solve the problems. Students could compile and illustrate a book of story problems and equations which they have created.

MANIPULATIVES: *Some problems may lend themselves to creating physical models to represent the equations students have created.*

SOLUTIONS

Answers will vary, but each solution, or story problem, should match the given equation. The value of the variable in each equation is given below.

1. C = 16   3. W = 5   5. M = 84
2. R = 4   4. A = 3

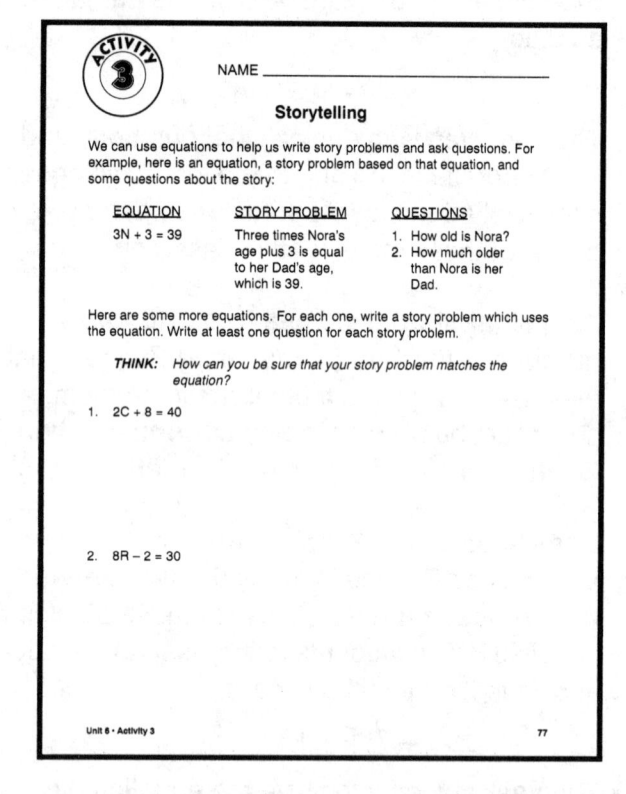

ACTIVITY 3

NAME _____

Storytelling

We can use equations to help us write story problems and ask questions. For example, here is an equation, a story problem based on that equation, and some questions about the story:

EQUATION	STORY PROBLEM	QUESTIONS
3N + 3 = 39	Three times Nora's age plus 3 is equal to her Dad's age, which is 39.	1. How old is Nora? 2. How much older than Nora is her Dad.

Here are some more equations. For each one, write a story problem which uses the equation. Write at least one question for each story problem.

THINK: How can you be sure that your story problem matches the equation?

1. 2C + 8 = 40

2. 8R − 2 = 30

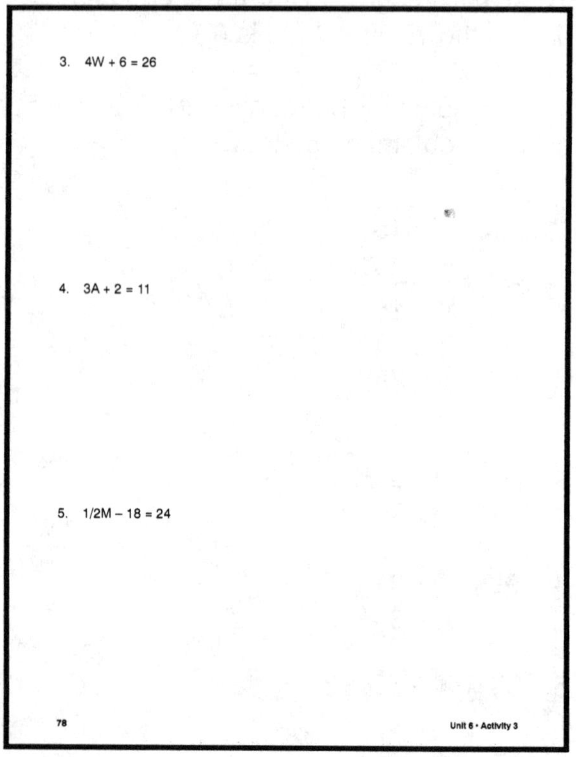

3. 4W + 6 = 26

4. 3A + 2 = 11

5. 1/2M − 18 = 24

---

74   Unit 6 • Activity 3

Activity 4

Formulas
Student Book, pages 79 – 80

FOCUS: Using formulas for area and perimeter
CONTENT CONNECTION: Geometry (MATHEMATICS)

In this activity, students will use equations to solve problems involving areas. They will also find the perimeter of irregular figures. In both situations, they will need to combine logical reasoning with their use of equations.

Note: You may want to have students reproduce the figures from this activity on centimeter graph paper.

For students who are not familiar with finding area and perimeter, you may want to review the definitions presented on student book page 79. You might also want to use examples such as the following:

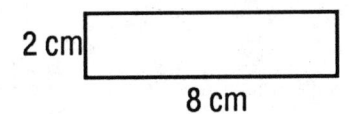

2 cm [rectangle]
8 cm

2 + 2 + 8 + 8 = 20 cm perimeter

2 × 8 = 16 cm² area

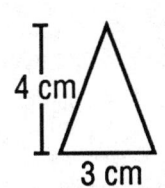

4 cm [triangle]
3 cm

1/2 bh = 1/2 × 3 × 4 = 6 cm² area

(2 cm) [circle]

πr^2 = 3.1416 × 4 = 12.57 cm²

For irregular figures (those which are not rectangles) like the ones in Problems 1 and 2, suggest that students first find the perimeters. Then they can *Solve a Simpler Problem* by finding the areas of the smaller rectangles which make up each figure, and *Work Backwards* to determine the area of the whole figure.

Problem 3 presents a *Guess and Check* problem in that students will have to decide how to design their figures. It will probably take several tries before they are successful.

Problems 4 and 5 also involve the *Solve a Simpler Problem* strategy. Students should first determine the area of the unshaded figures, and subtract that from the area of the rectangles.

CALCULATORS: Students will find calculators helpful, especially in problem 5 when they have to multiply by π (3.1416, approximately.)

In presenting their solutions to these problems, students should be prepared to explain, step-by-step, how they arrived at their conclusions.

Unit 6 • Activity 4

SOLUTIONS

1. Perimeter: 32 cm (A = 6 cm; B = 2 cm)
   Area: 48 cm²   4 x 10 = 40
                      2 x 4 = 8
                               ―――
                               48

2. Perimeter: 54 cm
   (9 + 6 + 1 + 5 + 1 + 4 + 9 + 4 + 2 + 5 + 2 + 6)
   Area: 120 cm²

   Note: Students can determine the area either by subtraction or by addition.

   Subtraction:
   Complete figure is   9 cm x 15 cm = 135 cm²
   Subtract: 2 cm x 5 cm = –10 cm²
                        1 cm x 5 cm = – 5 cm²
                                       ――――――
                                       120 cm²

   Addition:
   3 rectangles
             1. 9 cm x 6 cm = 54 cm²
             2. 6 cm x 5 cm = 30 cm²
             3. 9 cm x 4 cm = 36 cm²
                              ――――――
                              120 cm²

3. Answers will vary. Here is one possible solution:

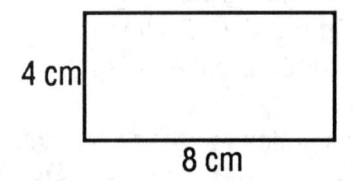

   4 cm
   [rectangle]
        8 cm

   perimeter = 24 cm (4 + 8 + 8 + 4)
   area = 32 cm² (4 x 8 = 32)

   Extension: Ask students to see how many different irregular figures they can construct with a perimeter of 24 cm and an area of 32 cm².

4. Shaded area = 52 cm²
   Rectangle   4 cm x 16 cm =   64 cm²
   Triangle   1/2 x 6 cm x 4 cm = – 12 cm²
                                    ――――――
                                     52 cm²

5. Shaded area = 94.16 cm²
   Rectangle   9 cm x 15 cm =   135 cm²   28.27 cm²   135.00 cm²
   Circle 1   3² x π = 9 x 3.1416 = – 28.27 cm²   + 12.57 cm²   – 40.84 cm²
   Circle 2   2² x π = 4 x 3.1416 = – 12.57 cm²   ――――――――   ――――――――
                                                     40.84 cm²   94.16 cm²

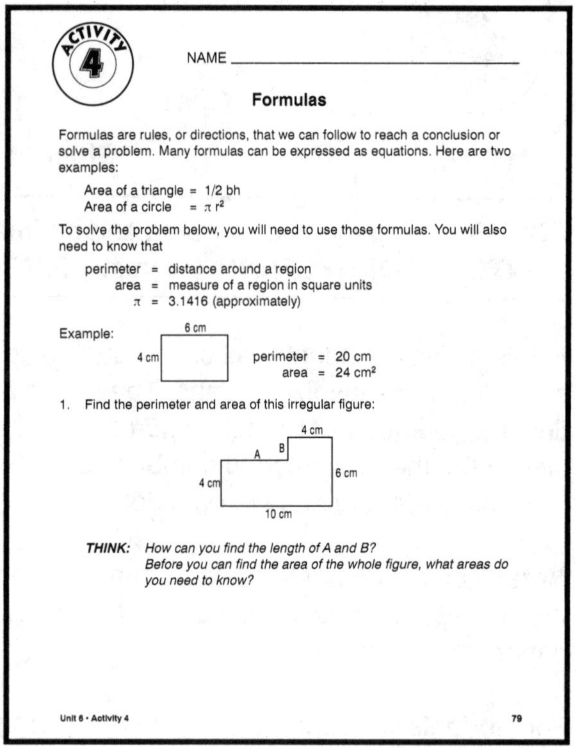

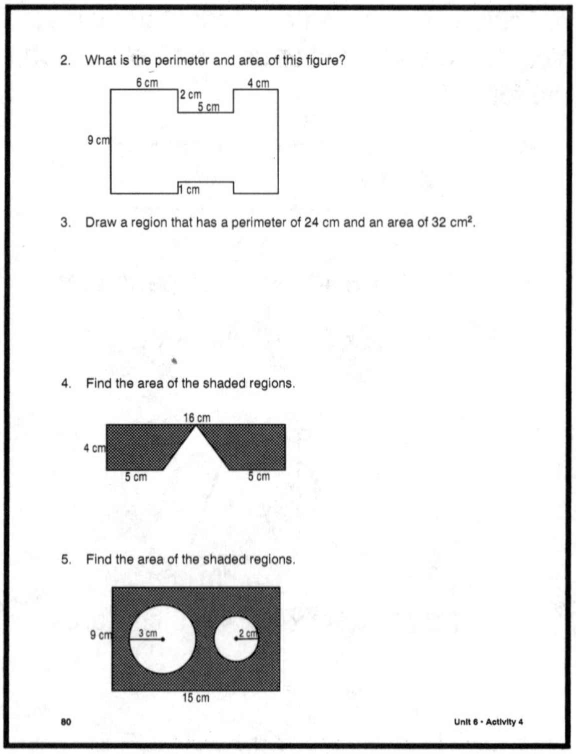

76   Unit 6 • Activity 4

ACTIVITY 5

Challenge
Student Book, pages 81 – 82

FOCUS: Review

These problems can serve as an assessment of students' understanding. They could also be assigned to small groups of students as additional practice. You may want to assign tasks to each group member to provide a cooperative learning group setting.

SOLUTIONS

1. 125 tickets for each
   5N − 13 = 612
   N = 125

2. Answers will vary.
   Problems should reflect the equation
   5N + 9 = 84, with N = 15.

3. Perimeter: 32 cm
   (5 + 8 + 2 + 2 + 2 + 2 + 2 + 5 + 1 + 3)
   Area: 41 cm²

Rectangle	5 cm x 8 cm =	40 cm²
Rectangle	1 cm x 5 cm =	+ 5 cm²
		45 cm²
Rectangle	2 cm x 2 cm =	− 4 cm²
		41 cm²

4. $8 each
   7N + 4 = 60
   N = 8

5. 37.7 cm²
   Larger circle π r² = π x 4² = π x 16 =   50.266 cm²
   Smaller circle π r² = π x 2² = π x 4 = −12.566 cm²
                                            37.7 cm²

Challenge

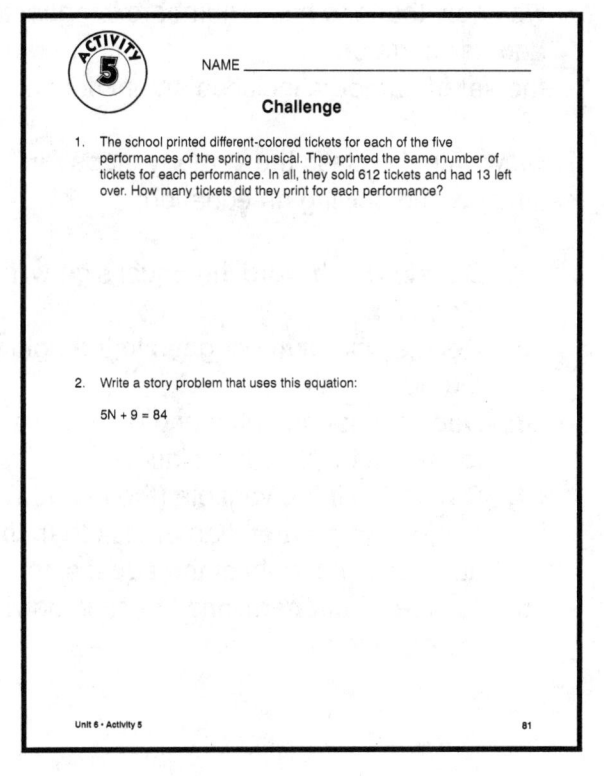

1. The school printed different-colored tickets for each of the five performances of the spring musical. They printed the same number of tickets for each performance. In all, they sold 612 tickets and had 13 left over. How many tickets did they print for each performance?

2. Write a story problem that uses this equation:
   5N + 9 = 84

3. Find the perimeter and area for this figure:

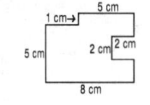

4. Cori and Carli had $60.00 to spend at the record store. They bought 7 compact discs and had $4.00 left. All the CD's were the same price. What was the cost of each CD?

5. Find the area of the shaded region.

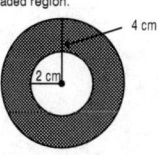

STRATEGY DISCUSSION

Point out to students that equations help us to solve certain kinds of problems. Ask them to list some of the characteristics of the problems in this unit. Help them to understand that equations work well for problems in which:

- we know that one set of numbers is equal to another number.
- the set of numbers includes an unknown.

Review with the class the strategy steps for setting up and solving an equation:

1. Decide which word the equal sign will stand for.
2. Decide what number goes to the right of the equal sign.
3. Decide what numbers and operations go to the left of the equal sign.
4. Determine if the variable (the unknown number) is greater than or less than the number to the right of the equal sign.
5. Decide what operations to use in solving the equation.

EXTENSION

Some students might like to create some perimeter, area, and volume problems. They could also make models of the equations they create. Encourage students to create a set of posters illustrating some of the equations they used.

Ask a team of students to create an original "Solve an Equation" poster for the classroom.

Students can pair up, as individuals or teams, to challenge each other. One side can create a problem; the other side finds an equation to solve it.

Interested students might like to turn back to the problems of earlier units to see if they can set up an equation for solving any of them.

Other students might like to give the perimeter and area of a figure and challenge other students to draw the figure.

Set Up an Equation

Student Book, pages 83 – 84

This page of the student text is designed to be taken home as a family involvement activity.

Included for the parent is a brief explanation of the unit content with a summary of the strategy which has been developed.

Along with that background information, there are several problems which the student and the family can solve together.

This page may be assigned as soon as Activity 4 of the unit has been completed.

SOLUTIONS

1. N = 15

2. 6 chairs

    4N + 3 = 27
    N = 6

3. He ordered 144 cans; 57 cans are left in the storeroom.

    $\frac{1}{2}$N + 15 = 87   144
    N = 144   − 87
                        57

4. Perimeter: 38 cm
    (4 + 2 + 2 + 6 + 2 + 2 + 2 + 5 + 2 + 11)
    Area: 46 cm^2
         Rectangle   2 cm x 11 cm =   22 cm^2
         Rectangle   2 cm x 6 cm =   12 cm^2
         Rectangle   2 cm x 6 cm = +12 cm^2
                                            46 cm^2

5. Shaded area = 64.868 cm^2
    Rectangle 6 cm x 18 cm   = 108 cm^2
    Triangle $\frac{1}{2}$ x 6 cm x 6 cm   = 18 cm^2
    Circle   2^2 x π = 12.566 x 2 = 25.132 cm^2

         18 cm^2 108 cm^2
    + 25.132 cm^2 − 43.132 cm^2
       43.132 cm^2 64.868 cm^2

Set Up an Equation

Recently, we have been learning how to use equations to solve problems.

Equations tell us that one set of numbers in a problem is equal to another number in the problem.

For example, in the equation 2N + 4 = 18, we know that two times some number, plus four, is equal to eighteen. The N stands for the unknown — the number we do not know and have to find.

In equations, letters usually stand for some unknown number. We might use N, or X, or any letter.

Sometimes we have to read the problem and then set up an equation. For example:

    Twice my age plus 4 equals my cousin's age, which is 18. How old am I?

We can turn that problem into the equation, 2N + 4 = 18. Once we set up the equation, we can solve it by finding what number N stands for. (N = 7)

Here are some problems like the ones we have been solving in school. You may enjoy solving them at home with your child.

1. Solve the equation:

    12N − 18 = 162

2. Four families each brought the same number of chairs to the block party. We still needed three more chairs to seat all 27 people who came to the party. How many chairs did each family bring?

3. A grocer ordered some cans of soup. He put half the cans in the storeroom, and half on the shelves. He sold so many during the day that he had to bring 15 more cans out from the storeroom. In all, by the end of the day, he had sold 87 cans.

    How many cans of soup did he order? How many were left in his storeroom?

4. Find the perimeter and area of this figure:

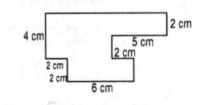

5. Find the area of the shaded region.

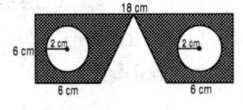

EXPERT PROBLEMS

Review Exercises

For students who wish to expand their experiences in problem solving, these exercises provide a review of the various strategies taught in each of the previous six units.

The exercises are designed to challenge the students first, to decide upon a strategy and second, to use that strategy to arrive at a solution.

Have students work with a partner or in small groups. Encourage students to discuss each problem, what strategy (or strategies) they would choose to solve it, and the reasons for choosing the strategy.

Once students have completed the problems, encourage a whole-class discussion with each group sharing the strategy they used to solve the problem, and giving their solution.

You may want to use these problems as an alternative assessment of students' understanding. Be sure students indicate *how* they solved each problem.

These exercises are cumulative, so that they can be used periodically throughout the year.

The following chart shows when each set of problems is appropriate:

Set	When to Use
Set A	After Unit 2
Set B	After Unit 3
Set C	After Unit 4
Set D	After Unit 5
Set E	After Unit 6

Note: With each solution, an appropriate strategy is listed. This is a strategy students *may* have used to solve the problem. However, any strategy students use that leads them to the correct solution is acceptable.

Student Book, pages 85 – 86

SOLUTIONS

1. 52 pillars

Strategy:
Estimate; Draw a Picture

2. The order is Adams, Monroe, Jefferson, Washington, and Madison.

Strategy:
Act It Out; Draw a Picture

3. It will look like this:

Strategy:
Draw a Picture; Make a Model

NAME _____

(Use after Unit 2.)

Use a strategy you have learned to solve the following problems.

1. The Governor's Mansion has a wide veranda around all four sides. If there are 14 white pillars on each side, how many pillars are there altogether?

   HINT: Use a drawing to check your solution.

2. In a hallway, the statues of the first five presidents have been moved around so that they are now out of order. Washington is between Madison and Jefferson. Three presidents, including Adams, are to the left of Washington as you face the statues. Neither Monroe nor Jefferson is at the end of the row. In what order — from left to right — are the statues?

   HINT: Use some of your classmates to represent the statues.

3. Rita has seven tiles that are 1 foot by 1 foot. She makes a design that has an area of 7 square feet and a perimeter of 14 feet. In her design, one tile touches just one other tile; another tile touches three other tiles; and all the rest touch two other tiles. ("Touch" means that sides touch completely, not just overlap.) What does Rita's design look like?

   HINT: Try drawing a picture to help you.

Expert Problems

Student Book, pages 87 – 88

SOLUTIONS

1. oldest, Mark; youngest, Latisha
   order: Mark, Adolfo, Karl, Esther, Fatima, Latisha

   Strategy:
   Make a Diagram

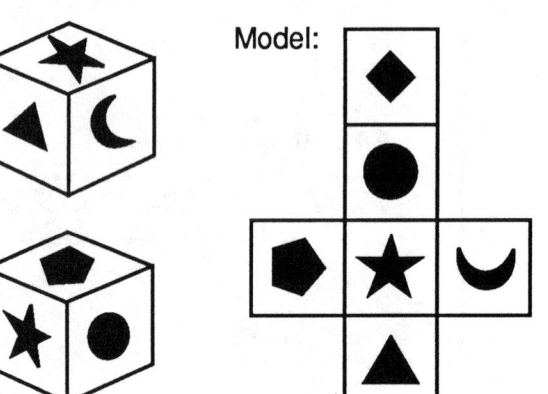

2. Model:

   Strategy:
   Make a Model

3. 55 tomatoes

	Tom	Pep	Cuke	Onion	TOTAL
Martha	21	5	3	2	31
Steve	15	0	4	3	22
Hilary	7	6	0	1	14
Sal	12	6	2	0	20
TOTAL	55	17	9	6	87

Strategy:
Make a Table

(Use after Unit 3.)

Use a strategy you have learned to solve the following problems.

1. Six friends were comparing their ages. Karl is older than Esther or Latisha. Only one person is older than Adolfo. Esther was born four months before Fatima. Latisha was born six months after Esther. Mark was born three weeks before Adolfo. Who is the oldest, and who is the youngest of the six?

   HINT: Put them in order on a time line from oldest to youngest.

2. Here are five views of the same cube. Fill in the blank spaces on the last two cubes.

   HINT: Make a model of the cube.

3. Martha, Steve, Hilary, and Sal volunteered to bring vegetables to make a salad for the Community Supper. Martha brought 21 tomatoes, 5 peppers, some cucumbers, and some onions. Steve brought 5/7 as many tomatoes as Martha, 4 cucumbers, and some onions — a total of 22 vegetables. Hilary brought no cucumbers, but supplied 14 vegetables in all (half of them were tomatoes). She brought six times as many peppers as onions. Sal brought 2 of the 9 cucumbers they used, and twice as many tomatoes as peppers. He did not bring any of the 6 onions they had. If they used 87 vegetables in all — along with a lot of lettuce — how many were tomatoes?

   HINT: Make a chart to help you keep track of the information.

Expert Problems

Student Book, pages 89 – 90

SOLUTIONS

1. 31 and 67

    Strategy:
    Guess and Check

2. 5:45

    Strategy:
    Solve a Simpler Problem (How long per room? — 15 minutes); Work Backwards

3. 12 more windows

Bruce	6	12	18	24	30	36	42	48
Kelly	7	14	21	28	35	42	49	56
Lena	5	10	15	20	25	30	35	40
TOTAL	18	36	54	72	90	108	126	144

    Strategy:
    Make a Table; Solve a Simpler Problem (How many windows? — 144) 48 – 36 = 12

4. It will take 7 trips:
    1. Take the monkey across.
    2. Go back alone.
    3. Take the fox (or the bananas) over.
    4. Bring the monkey back.
    5. Take the bananas (or the fox) over.
    6. Go back alone.
    7. Bring the monkey over.

    Strategy:
    Draw a Picture; Act It Out

	Left on Side 1		Left on Side 2
1.	fox bananas	monkey →	
2.	fox bananas	← alone	monkey
3.	bananas	fox →	monkey
4.	bananas	← monkey	fox
5.	monkey	bananas →	fox
6.	monkey	← alone	fox bananas
7.		monkey →	fox bananas

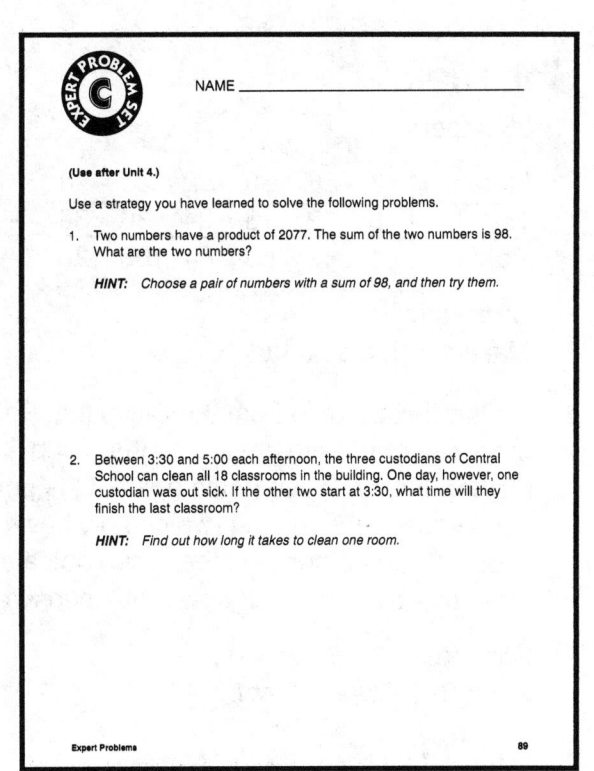

Expert Problems

SOLUTIONS

1. 36 papers

Florence	13	26	39	52	65	78	126
Ted	6	12	18	24	30	36	– 78
Ronnie	21	42	63	84	105	126	48

Strategy:
Make a Table or Chart

2. Fill the 2-cup bottle from the 8-cup jug. Pour the 2 cups from the bottle into the 3-cup pitcher. Fill the 2-cup bottle again. One person keeps the 8-cup jug (which now has 4 cups left); the other gets the 2-cup bottle and the 3-cup pitcher which now has 2 cups in it.

Strategy:
Act It Out; Draw a Picture

Do it in 3 steps:

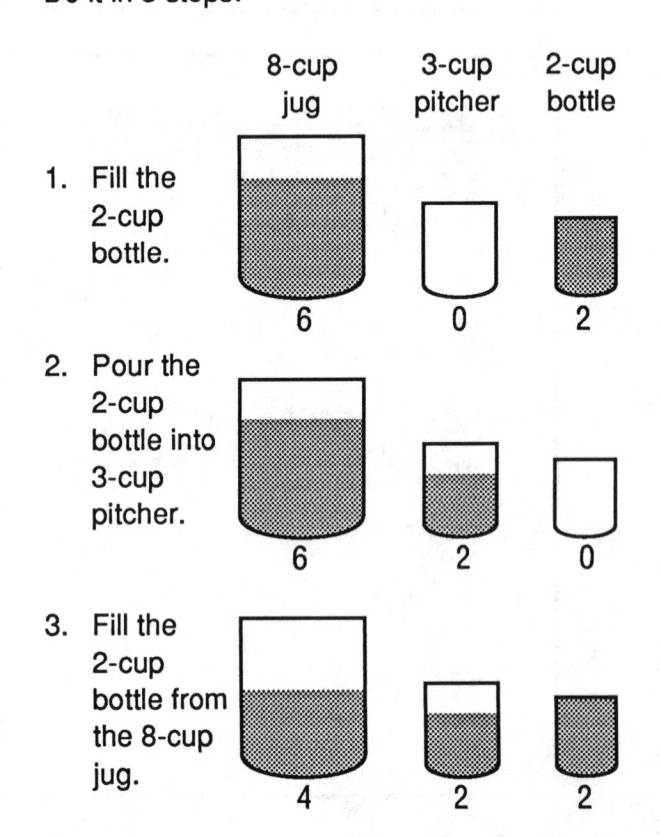

	8-cup jug	3-cup pitcher	2-cup bottle
1. Fill the 2-cup bottle.	6	0	2
2. Pour the 2-cup bottle into 3-cup pitcher.	6	2	0
3. Fill the 2-cup bottle from the 8-cup jug.	4	2	2

(Use after Unit 5.)

Use a strategy you have learned to solve the following problems.

1. For every 13 papers that Florence delivers, Ted delivers 6, and Ronnie delivers 21. When Ronnie has delivered 48 papers more than Florence, how many has Ted delivered?

   HINT: Can you set up a system to keep track of the numbers?

2. You need exactly 4 cups of molasses for a cookie recipe. Your neighbor is also baking cookies, and needs to borrow 4 cups of molasses. All you have to measure with is an 8-cup jug full of molasses, a 3-cup pitcher, and a 2-cup empty bottle. How can you and your neighbor each have exactly four cups of molasses?

   HINT: What do you pour into first?

3. 27 panes

Hour	1	2	3	4	5	6
Panes	12	15	18	21	24	27	= 117

Strategy:
Guess and Check; Make a Table

4. Coopers, Mexico, ocean beach; Kareems, Italy, lake; Pulvers, Canada, city; Chans, France, mountains.

	It	Can	Fr	Mex	mts	ocn	lake	city
Coopers	X⁴	X⁴	X⁴	•	X²	●	X²	X²
Chans	X¹	X¹	●	X⁴	●	X¹	X¹	X⁵
Kareems	●	X⁵	X²	X²	X⁵	X²	●	X⁵
Pulvers	X⁵	●	X²	X²	X³	X²	X⁵	●
mountains	X⁵	X³	●	X³				
ocean beach	X⁴	X⁴	X⁴	●				
lake	●	X⁵	X⁵	X⁵				
city	X⁵	●	X⁵	X⁴				

Clues:
1. Chans didn't go to Italy or Canada; don't like water, so didn't go to beach or lake.
2. Since Kareems and Pulvers didn't go to France, Mexico or an ocean beach, the Coopers, by elimination, must be the ones who went to a beach.
3. Mountains, not in North America, can't be in Canada or Mexico; Pulvers didn't go to mountains.
4. Ocean beach is in Mexico; since Coopers went to the beach (Clue 2), they went to Mexico; by elimination, the Chans went to France.
5. Pulvers didn't go to lake, so they must have gone to the city; lake is in Italy; therefore, by elimination, city where Pulvers went is in Canada, and the Chans, who went to France (Clue 4) must have gone to the mountains; the Kareems must have gone to a lake in Italy.

Strategy:
Use Logical Reasoning

Expert Problems   85

Student Book, pages 93 – 94

SOLUTIONS

1. 720 different ways

   3 objects = 6 ways
   4 objects = 4 x 6 = 24 ways
   5 objects = 5 x 24 = 120 ways
   6 objects = 6 x 120 = 720 ways
   7 objects = 7 x 720 = 5040 ways

   Strategy:
   Look for a Pattern

2. He bought the following:

$60.00		Power Saw	$19.95
– 5.88		Curtain Rod	6.49
$ 54.12 amount		Set of Wrenches	12.89
	spent	Pair of Hinges	+ 14.79
		TOTAL	$54.12

   Strategy:
   Solve a Simpler Problem; Guess and Check

3. Each bag held 24 balloons; 1 more bag needed.

   5N + 18 = 138
   N = 24

   Strategy:
   Set Up an Equation

4. Take one piece from the bag wrongly labeled "Peaches and Plums." Then, if it is a

Peaches/Plums	Peaches	Plums	Wrong Labels
peach	peaches	plums	peaches/plums	correct
plum	plums	peaches/plums	peaches	correct

   Strategy:
   Use Logical Reasoning; Make a Chart

5. It will take 7 hours.

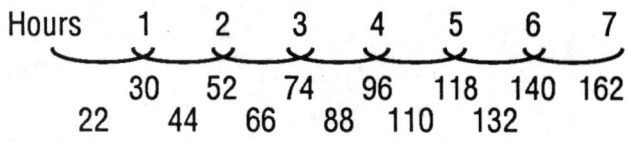

   Hours 1   2   3   4   5   6   7
          30 52 74 96 118 140 162
       22   44 66 88 110 132

86